Chemistry
IGCSE
Revision
Guide

RoseMarie Gallagher
Paul Ingram

OXFORD

OXFORD
UNIVERSITY PRESS

Great Clarendon Street, Oxford OX2 6DP

Oxford University Press is a department of the University of Oxford.
It furthers the University's objective of excellence in research, scholarship,
and education by publishing worldwide in

Oxford New York

Auckland Cape Town Dar es Salaam Hong Kong Karachi
Kuala Lumpur Madrid Melbourne Mexico City Nairobi
New Delhi Shanghai Taipei Toronto

With offices in

Argentina Austria Brazil Chile Czech Republic France Greece
Guatemala Hungary Italy Japan South Korea Poland Portugal
Singapore Switzerland Thailand Turkey Ukraine Vietnam

Oxford is a registered trade mark of Oxford University Press
in the UK and in certain other countries

British Library Cataloguing in Publication Data

Data available

ISBN 9780199152667

10 9 8 7 6 5 4

Printed in Great Britain by Bell and Bain Ltd., Glasgow

Cover photo: Tatiana53/Dreamstime

Paper used in the production of this book is a natural, recyclable
product made from wood grown in sustainable forests. The
manufacturing process conforms to the environmental regulations of
the country of origin.

CIE past paper examination material reproduced by permission of the University of Cambridge
Local Examinations Syndicate.

The University of Cambridge Local Examinations Syndicate bears no responsibility for the
example answers to questions taken from its past question papers which are contained in this
publication.

Contents

About this book

What is it for?

The aim of this book is to help you succeed in your Cambridge IGCSE chemistry exams.

The book covers the Cambridge IGCSE Chemistry syllabus 0620. It is divided into 22 main sections, which follow the order of the topics in the syllabus.

Material for the Extended curriculum *only* is indicated by a grey wavy line. This material is not required if you are studying the Core curriculum. If you are studying the Extended curriculum, *all* of the material in the book applies to you.

The questions

The 'Quick check' questions at the end of each subsection are to check that you understand the key points it covers.

At the end of each section, you will find a set of exam and exam-level questions, on the topics in that section. These will prepare you for the kinds of questions that appear in the written exams.

The exam questions are from Cambridge IGCSE past papers. They include questions from paper 2 (Core), paper 3 (Extended) and paper 6 (Alternative to practical). The references in brackets tell you the year, session, and paper.

Questions from paper 6 apply to both Core and Extended candidates. Even if your school does not take paper 6, you will still find it useful to work through these questions. They will help you with other papers, including the practical exam.

Answers for all questions are provided at the back of the book.

Using the book

Revision plays a big part in exam success. So …

- leave yourself plenty of time to revise. Don't wait until the last minute!
- plan which section(s) of the book to cover in each revision session. This will help you feel in control.
- remember that several short revision sessions can be more effective than one long one. Your brain needs a rest!
- leave time for working through the exam and exam-level questions. Answering questions is a very good way to revise.
- you may find that it helps to make your own notes from the book as you go along.

We hope that this book will help you do well in your Cambridge IGCSE chemistry exams.

1 Substances are made of particles

The big picture

- All the substances around us are made of tiny pieces or **particles**.
  This idea is called **the particle theory**.
- Substances can be solid, liquid, or gas, and can change from one state to another.
  For example, ice can change to water.
- Solids, liquids, and gases have different properties.
- We can explain the differences by looking at how their particles are arranged.

1.1 The particle theory

What is the particle theory?

- The **particle theory** says that all matter is made of very tiny pieces or **particles**.
  (As you will see in Section 4, these particles can be single atoms, or molecules, or ions.)
- The particles are arranged differently in a solid, liquid, and gas – and that is why these
  states have different properties.

The particle arrangement in a solid, liquid, and gas

State of the substance	solid	liquid	gas
	example: ice	example: water	example: water vapour
The particle arrangement			
The particles are …	close together	still close together	far apart
The arrangement is …	regular	irregular	random
The forces between the particles are …	strong or quite strong	less strong	non-existent
… so the particles can …	vibrate to and fro, but not move apart; that is why a solid has a fixed shape	move around, and slide past each other; that is why a liquid can be poured	move freely, collide with each other, and bounce away again; that is why a gas spreads
When you apply pressure …	the particles can't move closer, so the volume does not change	the particles can't move closer, so the volume does not change	the particles can move a lot closer, so the gas can be pushed into a much smaller volume

Some evidence for particles

How do we know there are particles, and that they can move? Here is some evidence:

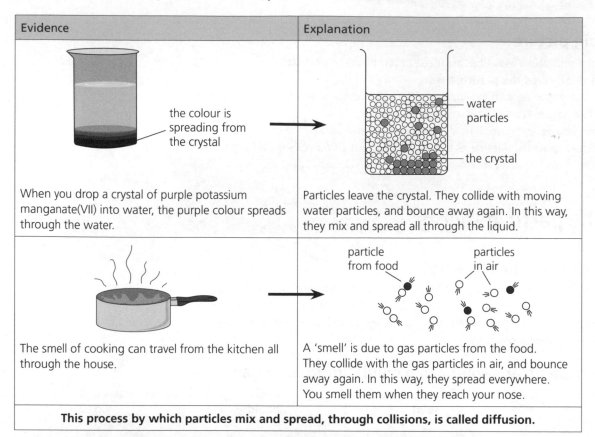

Evidence	Explanation
When you drop a crystal of purple potassium manganate(VII) into water, the purple colour spreads through the water.	Particles leave the crystal. They collide with moving water particles, and bounce away again. In this way, they mix and spread all through the liquid.
The smell of cooking can travel from the kitchen all through the house.	A 'smell' is due to gas particles from the food. They collide with the gas particles in air, and bounce away again. In this way, they spread everywhere. You smell them when they reach your nose.
This process by which particles mix and spread, through collisions, is called diffusion.	

✓

Quick check for 1.1 *(Answers on page 164)*

1 Using the particle theory, explain why:
   a you can pour water   **b** you can not squeeze water into a smaller volume.

2 You can pump a great deal of air into a bike tyre. Why?

3 You put some sugar cubes in a glass of cold water, without stirring. After a time, all the water tastes sweet. Name, and explain, the process that caused this.

1.2 Changing state

When you heat a solid

You can change the state of a substance from solid to liquid to gas, by heating it. The particles take in **heat energy** and change it to **kinetic energy** (or energy of movement):

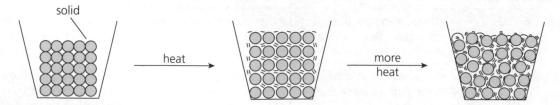

1 In a solid, the particles are held in a regular structure. They don't move away, but they do vibrate to and fro.

2 As the particles take in heat energy, the vibrations get larger and stronger. **So the solid expands a little.**

3 With more heat, the particles vibrate so much that the structure breaks down. **The solid melts to a liquid.**

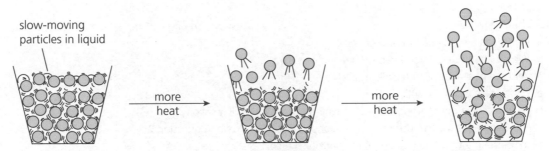

slow-moving
particles in liquid

more
heat

more
heat

4 The particles continue to take in heat energy. So they move around more. **So the liquid expands a little**.

5 Some particles gain enough energy to overcome the forces between them, and escape. **This is evaporation**.

6 At a certain point, all the remaining particles gain enough energy to escape. **The liquid boils to a gas**.

Showing the changes on a graph

Suppose you apply steady heat to a block of ice. You record how its temperature changes with time. A graph of your results is called a **heating curve**. It will look like this:

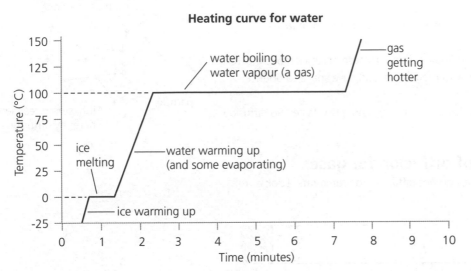

Heating curve for water

water boiling to
water vapour (a gas)

gas
getting
hotter

ice
melting

water warming up
(and some evaporating)

ice warming up

Temperature (°C)

Time (minutes)

> **Remember**
> When a substance changes state the particles do not change – only their arrangement.

- Note that the temperature stays steady at 0 °C (the melting point), until all the ice has melted to water.
- The temperature stays steady at 100 °C (the boiling point) until all the water has turned to a gas (water vapour).
- If you cool the hot water vapour down, your graph will be a **cooling curve**. It will be a mirror image of the heating curve above.

> **Note**
> Heating curves for other substances are a similar shape: flat at their melting and boiling points.

Reversing the changes

Those changes of state can be reversed by cooling. Look at this:

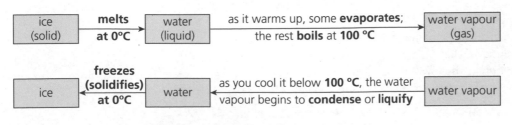

On heating, the particles gain energy and move faster; in time they gain enough energy to overcome the forces of attraction between them.

| ice (solid) | **melts** **at 0°C** | water (liquid) | as it warms up, some **evaporates**; the rest **boils** at **100 °C** | water vapour (gas) |

| ice | **freezes** **(solidifies)** **at 0°C** | water | as you cool it below **100 °C**, the water vapour begins to **condense** or **liquify** | water vapour |

On cooling, the particles lose energy and move more slowly; as they get closer together, the forces of attraction take over.

(Answers on page 164)

✓

Quick check for 1.2

1 **a** Why does a solid expand on heating?
   b What will happen if you keep on heating it? Why?
2 A solid melts at 130 °C and boils at 220 °C.
   a What state is it in at 240 °C?
   b As you cool it down from 240 °C, at what temperature will it solidify?
3 Sketch the heating curve obtained, if you apply heat to the solid in question **2**. Mark in the melting and boiling points on your sketch.

1.3 More about diffusion

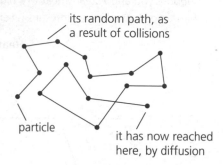

- Diffusion is the process by which particles mix and spread, through collisions with other particles.
- It is a random process. Look at this diagram. The path a particle takes depends on its collisions.
- It is much faster in gases than in liquids. That's because particles move much faster in gases. So they collide with more force, and have space to bounce further away.
- As the temperature rises, particles take in energy and move faster. So diffusion is faster too.

Comparing the rates of diffusion for gases

Extended

Even at the same temperature, gases do not diffuse at the same rate. Look at this experiment:

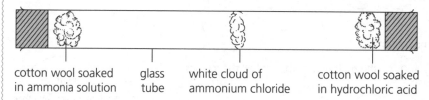

cotton wool soaked
in ammonia solution

glass
tube

white cloud of
ammonium chloride

cotton wool soaked
in hydrochloric acid

- Particles of ammonia gas and hydrogen chloride gas diffuse from the opposite ends of the long glass tube. (The particles are molecules.)
- When they meet, they combine to form a white cloud of ammonium chloride.
- The white cloud forms closer to the right-hand end. So the ammonia molecules have travelled faster. That's because they are lighter. (The relative molecular masses are: ammonia 17; hydrogen chloride 36.5.)

The lower its relative molecular mass, the faster a gas will diffuse.

✓

Quick check for 1.3

(Answers on page 164)

1 The experiment above uses ammonia and hydrogen chloride gases. What makes these a good choice?
2 Chlorine and nitrogen are made of molecules. Their relative molecular masses are: chlorine, 71; nitrogen, 28.
   a Which gas will diffuse more slowly at room temperature? Why?
   b What could you do, to try to make it diffuse *faster* than the other gas?

Questions on section 1

Answers for these questions are on page 164.

Core curriculum

1   The states of matter are solid, liquid and gas.
    a   Draw diagrams to compare the arrangement of the particles in a solid and a liquid.
    b   Describe the movement of particles in
        i   the solid state
        ii   the liquid state
    c   Describe what happens to the particles in a gas when it condenses to form a liquid.
    d   Which state contains particles with the greatest kinetic energy?
    e   Why does a gas expand on heating?

2   A student placed a crystal of copper(II) sulfate in a beaker of water.
    After one hour the crystal had completely disappeared and a dense blue colour was
    observed in the water at the bottom of the beaker. After 48 hours the blue colour had
    spread throughout the water.

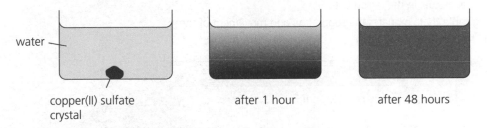

water

copper(II) sulfate   after 1 hour   after 48 hours
crystal

    a   Use the kinetic particle theory to explain these observations.
    b   For the copper(II) sulfate **crystal** describe:
        i   the arrangement of the particles
        ii   the motion of the particles   *CIE 0620 June '08 Paper 2 Q7*

3   The long glass tube was set up as shown in this diffusion experiment.

            glass tube   rubber bung

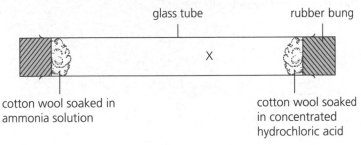

                                X

    cotton wool soaked in   cotton wool soaked
    ammonia solution   in concentrated
                                       hydrochloric acid

    a   Which gases are present in the tube?
    b   A cloud of a substance will form around X in the tube.
        i   Name the substance.
        ii   What colour will the cloud be?
    c   The substance that forms around X is evidence that gas particles are spreading
        in the tube. Explain why.
    d   What causes the gas particles to spread?
    e   What will happen if the cotton wool soaked in hydrochloric acid is replaced by
        cotton wool soaked in ammonia?

Extended curriculum

1 Ethanoic acid is a colourless liquid at room temperature.

   a A pure sample of ethanoic acid is slowly heated from 0 °C to 150 °C and its temperature is measured every minute. The results are represented on this graph:

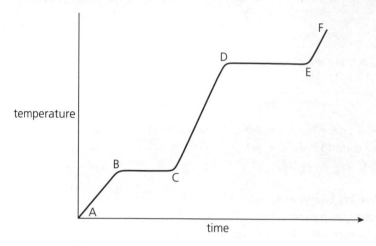

     i Name the change that occurs in the region B to C.

     ii Name the change that occurs in the region D to E.

   b Complete the following table that compares the separation and movement of the molecules in regions C to D with those in E to F.

	C to D	E to F
separation (distance between particles)	**i**..................................	**ii**..................................
movement of particles	random and slow	**iii**..................................
Can particles move apart to fill any volume?	**iv**..................................	**v**..................................

CIE 0620 November '05 Paper 3 Q2a i and iv

2 This table shows the formulae and relative molecular masses for four elements that are gases at room temperature.

Gas	Formula	Relative molecular mass
chlorine	Cl_2	71
hydrogen	H_2	2
helium	He	4
oxygen	O_2	32

   a **i** Draw a diagram to show the arrangement of particles in a gas.

     ii What can you say about the forces between molecules, in the gas state?

     iii Describe the movement of the molecules, in the above gases.

   b **i** Arrange the four elements in order of the speed at which they will diffuse in air (quickest first).

     ii Explain how you made this choice.

   c Compare the rate of diffusion of methane CH_4 (relative molecular mass of 16) with oxygen.

   d An unknown gas X diffuses more slowly than oxygen, but faster than chlorine. What can you say about the molecular mass of X?

2 Experimental techniques

The big picture

- Our knowledge of chemistry is based on experiments.
- Experiments involve taking measurements. For example, weighing things.
- Chemists have developed many different techniques to use in experiments.
- These include techniques for separating substances from mixtures, and checking that substances are pure.

2.1 Taking measurements in the lab

When you do experiments, you usually have to measure something. Look at this table.
Pay special attention to the units used!

What is measured?	Units of measurement	Apparatus
time	seconds (s) minutes (min) **60 s = 1 min**	stopclock
temperature	degrees Celcius (°C)	thermometer
mass	grams (g) kilograms (kg) **1000 g = 1 kg**	weighing balance
volume of liquid	cubic centimetre (cm³) or millilitre (ml) cubic decimetre (dm³) or litre (l) Make sure you know that: **1 cm³ = 1 ml** **1 dm³ = 1 l** **1000 cm³ = 1 dm³ = 1 l**	increasing accuracy of measurement measuring cylinder burette pipette

Note

- A stopwatch could be used instead of a stopclock, for measuring time.
- There are many different kinds of weighing balance.
- Your lab may have measuring cylinders, burettes, and pipettes in different sizes.
  For example pipettes to measure out 10.0 cm³, and 25.0 cm³.

✓ **Quick check for 2.1** (Answers on page 164)

1   You have to measure out 3 g of sodium chloride and dissolve it in approximately 100 cm^3 of water. What apparatus will you use?
2   What would you use in the lab, to measure out very accurately 25.0 cm^3 of a solution?
3   Give this in grams: **a** 2.5 kg   **b** 0.5 kg
4   Give this in cm^3:   **a** 0.5 l   **b** 2 dm^3
5   Give this in dm^3:   **a** 2 l   **b** 0.5 l   **c** 1500 cm^3 **d** 500 cm^3

2.2 Pure and impure substances

When you make something in the lab, it is not usually pure. It may have other things mixed with it. For example water, or a reactant that has not reacted.

A pure substance ...
- has no particles of any other substance mixed with it
- melts and boils at temperatures that are unique to it; for example ice melts at 0 °C and water boils at 100 °C; no other substance has those melting and boiling points
- melts sharply at its melting point

An impure substance ...
- has particles of another substance mixed with it
- melts over a range of temperature, not sharply
- –   melts at a lower temperature than the pure substance
  –   boils at a higher temperature than the pure substance.

So:
- by measuring the melting and boiling point of a substance, you can tell if it is pure.
- you can identify an unknown pure substance, by measuring its melting and boiling points, and then looking these up in tables.

It is essential to make sure some substances are pure. For example substances used as food additives, and in medical drugs and vaccines.

✓ **Quick check for 2.2** (Answers on page 164)

1   Pure ethanol boils at 78.4 °C and freezes at –114.3 °C. You have a sample of ethanol which boils between 79.1 °C and 79.9 °C.
    a What can you say about its purity?
    b What will you expect to find, when you measure its freezing point?
2   It is very important that some substances are pure. Give two examples.

2.3 Ways to purify a substance

To purify a substance, you have to separate the unwanted substances from it.
Before looking at separation methods, make sure you understand these terms:

mixture – two or more substances that are not chemically combined
solution – a mixture you make by dissolving a substance in a solvent
solute – the substance that you dissolve
solvent – the liquid in which you dissolve the substance

Separation methods

This table shows four separation methods.

Filtration	Crystallisation
To separate: an insoluble solid from a liquid	**To separate:** the solute from a solution

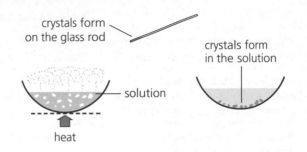

Filtration

To separate: an insoluble solid from a liquid

filter paper
filter funnel
mixture
solid trapped
flask
liquid (filtrate)

- The solid remains on the paper. You can rinse it with distilled water.
- The liquid or **filtrate** passes through the paper.

Crystallisation

To separate: the solute from a solution

crystals form on the glass rod
crystals form in the solution
solution
heat

Crystallisation works because a solvent can dissolve less and less solute, as its temperature falls. (A **saturated solution** can dissolve no more solute, at that temperature.)
- Heat the solution, to evaporate some solvent.
- Test the hot solution using a glass rod. If crystals form on the rod, you know the solution is saturated.
- If the solution is saturated, stop heating, and leave it to cool. As it cools, crystals of the solute form.
- Separate the crystals by filtering. Wash them with a little distilled water. (Take care, since they are soluble.)

Simple distillation

To separate: the solvent from a solution

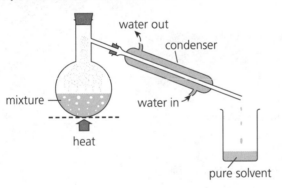

water out
condenser
mixture
water in
heat
pure solvent

- The solvent boils off as a gas, at its boiling point.
- The gas condenses back to a liquid in the cool condenser.
- The solute remains in the flask.

Fractional distillation

To separate: two or more miscible liquids (liquids that mix completely)

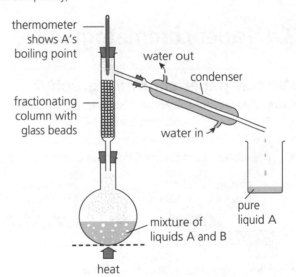

thermometer shows A's boiling point
water out
condenser
fractionating column with glass beads
water in
mixture of liquids A and B
pure liquid A
heat

Suppose A has the lower boiling point.
- Gas A passes into the condenser first, at A's boiling point. It condenses there.
- Collect liquid A as above. Stop collecting when the temperature on the thermometer rises.
- Keep heating. Gas B passes into the condenser, at B's boiling point. Collect liquid B in a separate beaker.

Examples of separations

Those separation methods are used in the laboratory and in industry, as this table shows.

Mixture	How to separate and purify the substance you want
Sand and salt.	1 Add water to dissolve the salt; the sand will not dissolve. 2 Filter to remove the sand. 3 Evaporate the filtrate to obtain pure salt.
Alcohol, water, and other substances, from the fermentation of grapes.	Use fractional distillation to distil the alcohol, which boils at a lower temperature than water. (See page 148.)
Petroleum (crude oil) which is a mixture of hundreds of compounds, all with different boiling points.	Use fractional distillation to separate the compounds into groups, each with quite a small range of boiling points. Collect the groups in order of their boiling points. (See page 140.)
Air, which is a mixture of gases (mainly nitrogen and oxygen).	Cool the air until it is liquid, then use fractional distillation to boil the gases off one by one. (See page 125.)
Reaction mixture for making aspirin.	Cool the reaction mixture in ice. Pure aspirin crystallises out as the mixture cools. Wash the aspirin crystals with distilled water to remove impurities.

✓

Quick check for 2.3   (Answers on page 164)
1   Explain what these terms mean: **a** solute   **b** solvent
2   Compare the apparatus for simple distillation and fractional distillation. What differences do you notice?
3   Give an example of when to use: **a** simple distillation   **b** fractional distillation
4   Explain *why* crystallisation works, as a way to purify a substance.

2.4 Paper chromatography

What is paper chromatography?

Paper chromatography is another way to separate substances from a mixture.

•   The mixture is dissolved in a suitable solvent.
•   The solution is allowed to travel across paper.
•   The substances in the solution travel at different speeds. So they separate, like this:

mixture of substances   they travel at   so they separate
dissolved in a solvent   different speeds

paper

•   They separate because of their different solubility in the solvent, and attraction to the paper they travel over.
•   The more soluble a substance is in the solvent, the further it will travel. The more attracted it is to the paper, the less far it will travel.

What is paper chromatography used for?

Paper chromatography is used:

➤   to find out how many substances are present, in a mixture
➤   to check on the purity of a substance; if impurities are present, they separate out
➤   to identify the substances in a mixture, by measuring how far they travel.

Look at the two examples that follow.

Example 1: separating the dyes in ink

1	2	3	4
Mark a base line on the paper, in pencil. (Biro or marker pen might run!) Place a spot of the ink on it.	Stand the paper upright in a beaker containing a little water (the solvent). Put a lid on the beaker.	Allow the water to rise up the paper. When it is near the top, remove the paper, and dry it in an oven.	The **chromatogram** (paper plus spots) shows that this black ink is made up of three dyes: red, yellow and blue.

base line marked in pencil

chromatography paper

solvent

yellow spot
red spot
blue spot

Example 2: separating colourless substances

You make them show up by using a **locating agent**. Look at this example.

1	2	3	4
Put a spot of a colourless solution of amino acids on a base line (marked in pencil). Stand the paper in solvent.	Remove the paper, and mark a line (in pencil, as before) to show where the solvent has reached.	Dry the paper out in an oven, then spray it with a locating agent. (Ninhydrin is a suitable one.)	Measure from the base line to the centre of each dot (see A below) and to the solvent front (B).

solvent front

A   B

A mixture of butan-1-ol, ethanoic acid and water is a suitable solvent.	But you will not see any spots, because amino acids are colourless.	Three purple spots appear – so the solution contained three amino acids.	To **identify** each amino acid, work out its R_f value: $$R_f = \frac{A}{B}$$

R_f value $= \dfrac{\text{distance moved by amino acid}}{\text{distance moved by solvent}}$

Each amino acid has a specific R_f value in a given solvent. You can look the values up in tables. For example, for the solvent above, the R_f values are:
0.26 for glycine
0.73 for leucine.

✓ **Quick check for 2.4**   (Answers on page 164)

1 Explain *why* chromatography can be used to separate a mixture of dyes.
2 In Example 1 above, which dye is the most soluble in water?
3 Explain how chromatography can be used to **identify** an amino acid.
4 How will a pure amino acid show up on a chromatogram?
5 On a chromatogram, the solvent reached 8 cm above the base line. The solute spot is 4.5 cm above the base line. Calculate the R_f value for the solute.

Questions on section 2

Answers for these questions are on page 164.

Core curriculum

1   A list of techniques used to separate mixtures is given below.

fractional distillation   simple distillation   crystallisation   filtration   diffusion

From the list choose the most suitable technique to separate the following.
a   water from aqueous copper(II) sulfate
b   helium from a mixture of helium and argon
c   copper(II) sulfate from aqueous copper(II) sulfate
d   ethanol from aqueous ethanol
e   barium sulfate from a mixture of water and barium sulfate
(barium sulfate is insoluble in water)

CIE 0620 November '07 Paper 3 Q1

2   **a**   The diagram shows how natural mineral water is formed from rainwater (surface
water) that soaks through limestone. Mineral water contains **no bacteria** or
particles of earth.

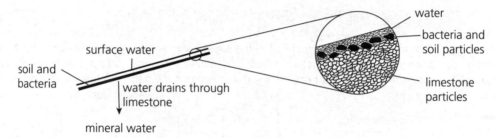

Use the diagram to explain how the mineral water is purified from bacteria and
particles of earth.
    b   Pure water can be obtained by distilling mineral water using the apparatus
shown below.

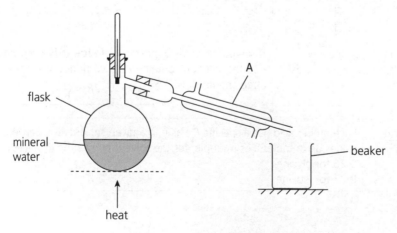

i   State the name of the piece of apparatus labelled A.
ii   Where does the pure water collect?
iii   How does the boiling point of the mineral water in the flask compare
with the boiling point of pure water?

CIE 0620 November '07 Paper 2 Q2e and f

3 The diagram shows the salt mines at Bex in Switzerland.

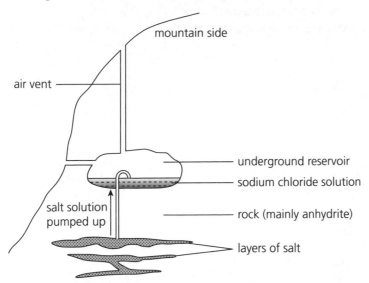

The salt is dissolved by water from underground springs and then pumped up to a reservoir where it is stored as a solution.

a Is sodium chloride soluble, or insoluble, in water?

b Suggest how solid sodium chloride is obtained from the sodium chloride solution.

c What separation process could be used to allow the water to be recovered?

d Is anhydrite (the rock surrounding the layers of salt) soluble or insoluble in water?

CIE 0620 November '05 Paper 2 Q2

4 A metal coin is dissolved in acid. Chromatography is used to test the solution formed. The diagram on the right shows the chromatogram obtained.

a Describe how the chromatogram would be set up in the laboratory.

b What can you say about the composition of the coin?

c Which of the spots (A or B) is **more** soluble in the solvent that was used in the chromatography?

d Which of the spots (A or B) is **more** attracted to the chromatography paper?

Extended curriculum

1 A mixture of two amino acids was separated by chromatography. The results are shown below.

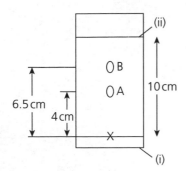

a Which of the two lines, (i) or (ii), shows where the solvent reached?

b Which amino acid is more soluble in the solvent that was used?

c The amino acids are colourless. How do you find where they are, on the paper?

d Calculate the R_f values for the two amino acids, A and B.

e Suggest how the amino acids could be identified.

f The R_f values will be different, for a different solvent. Explain why.

Alternative to practical

1 A mixture of ethanol and water can be separated by fractional distillation.
The apparatus below can be used to carry out such a separation in the laboratory.

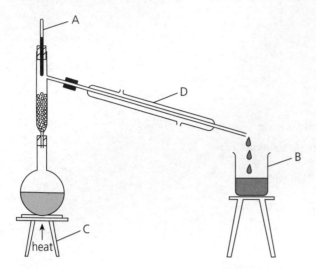

    a   Name each piece of apparatus, labelled A to D.

    b   What is the purpose of D?

    c   How could the purity of the ethanol collected be checked?

CIE 0620 June '07 Paper 6 Q1

2 Chromatography can be used to identify amino acids from a sample of protein.
The diagram shows the chromatogram obtained when four samples of amino acids
were analysed. The paper was sprayed with ninhydrin.

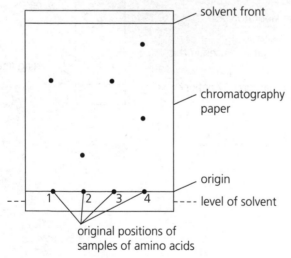

    a   Why is the origin line drawn in pencil?

    b   Which amino acids could possibly be the same?

    c   Which amino acid sample contains more than one amino acid?
        Explain your answer.

    d   Suggest why it is necessary to spray the chromatogram with ninhydrin.

CIE 0620 June '07 Paper 6 Q3

3 Atomic structure and the Periodic Table

The big picture

- Elements consist of tiny units called atoms.
- Atoms are made up of smaller particles: protons, electrons and neutrons.
- The protons dictate which element the atom belongs to.
- The electrons dictate how the element reacts.

3.1 The structure of the atom

Protons, electrons and neutrons

- Elements consist of tiny units called atoms.
- Atom are themselves made up of smaller particles: **protons**, **electrons**, and **neutrons**.
- These particles have so little mass that it is not given in grams. It is given in special units called **relative atomic mass units**.

Note

Because they make up the atom, protons, neutrons and electrons are called subatomic particles.

This table shows the properties of the particles:

Particle	Mass, in relative atomic mass units	Charge
proton (p)	1	1 + (positive)
neutron (n)	1	no charge
electron (e)	1/1840 (or almost nothing, so it is usually ignored)	1 − (negative)

Where they are, in an atom

Let's take a magnesium atom as example. It has 12 protons, 12 electrons and 12 neutrons. Here is a drawing of it:

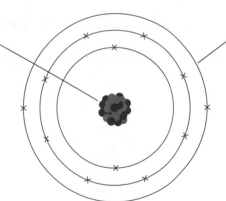

1 The protons and neutrons form a **nucleus** in the centre of the atom. So they are also called **nucleons**.

2 The **nucleon number** is the total number of protons and neutrons. So for magnesium it is 24.

3 The electrons circle very fast around the nucleus, at different energy levels from it. (The energy levels are called **shells**.)

4 Electrons have negligible mass. So the atom's mass is due to its nucleus. The nucleon number tells us the relative atomic mass of the atom. So it is also called the **mass number**.

More about protons

- The number of protons in an atom tells you which **element** it belongs to. If an atom has 12 protons, it is a magnesium atom. (If it has 13, it is an aluminium atom.)
- It also tells you how many **electrons** there are. An atom has no overall charge, so the + and − charges must balance each other. Therefore:
  number of electrons = number of protons
- The number of protons is also called the **proton number**.
- The elements in the Periodic Table are in order of their proton number. (See page 105.)

Shorthand for an atom

You can describe the atoms of an element in a short way, like this:

nucleon number —— A
proton number —— Z **X** —— symbol for the element

$^{A}_{Z}X$

A (nucleon number)	= number of protons + number of neutrons
Z (proton number)	= number of protons = number of electrons
So number of neutrons	= A – Z

So for magnesium, the shorthand is: $^{24}_{12}Mg$

This tells you that the magnesium atom contains 12 protons, 12 electrons, and 12 neutrons. (24 – 12 = 12). So its relative atomic mass is 24.

✓

Quick check for 3.1   *(Answers on page 164)*

1 Which of the particles in an atom has:
   a no charge?   **b** negligible mass?   **c** a positive charge?   **d** a charge of 1–?

2 An atom of lithium has 3 protons. Its nucleon number is 7.
   a How many electrons does it have?
   b How many neutrons does it have?
   c What is the mass number, for lithium atoms?
   d Describe the lithium atom in the short way, using its symbol and numbers.

3 Say as much as you can about these atoms: **a** $^{23}_{11}Na$ **b** $^{40}_{20}Ca$ **c** $^{56}_{26}Fe$

3.2 Isotopes

What are isotopes?

- The atoms of an element are not always identical.
- They all have the same number of protons and electrons. But *they can have different numbers of neutrons*.
- Atoms of the same element, with different numbers of neutrons, are called **isotopes**.

For example magnesium has three isotopes. They all occur naturally. Out of every 100 magnesium atoms ....

… about 80 of them, or 80% are:	… about 10 of them, or 10% are:	… and about 10% are:
$^{24}_{12}Mg$	$^{25}_{12}Mg$	$^{26}_{12}Mg$
(or magnesium-24 for short)	(or magnesium-25 for short)	(or magnesium-26 for short)
This isotope contains: 12 protons 12 electrons 24 – 12 = **12 neutrons**	This isotope contains: 12 protons 12 electrons 25 – 12 = **13 neutrons**	This isotope contains: 12 protons 12 electrons 26 – 12 = **14 neutrons**

- So each isotope has a different relative atomic mass. Their *average* relative atomic mass is 24.3.
- But the isotopes are *chemically* similar, because they have the same number of electrons.

How to calculate the average relative atomic mass of the isotopes

The average relative atomic mass of the isotopes =
% × mass for first isotope + % × mass for second isotope + % × mass for third isotope and so on.

So the average relative atomic mass of the isotopes for magnesium =

$$\left(\frac{80}{100} \times 24\right) + \left(\frac{10}{100} \times 25\right) + \left(\frac{10}{100} \times 26\right) = 19.2 + 2.5 + 2.6 = \mathbf{24.3}$$

Note that it is usually rounded off to **24**, for calculations.

Radioactive isotopes

- Some isotopes are unstable. They break down naturally, or **decay**, over time, giving out **radiation**.
- So these isotopes are called **radioactive**.
- A radioactive isotope is often called a **radioisotope**.

For example, sodium has three isotopes. Two of them are two unstable (radioactive).

Stable (non-radioactive)	Unstable (radioactive)	Unstable (radioactive)
$^{23}_{11}\text{Na}$	$^{22}_{11}\text{Na}$	$^{24}_{11}\text{Na}$
(or sodium-23)	(or sodium-22)	(or sodium-24)
This isotope has **12 neutrons** (= 23 − 11)	This isotope has **11 neutrons** (= 22 − 11)	This isotope has **13 neutrons** (= 24 − 11)

More about radiation

Radiation consists of tiny particles, and rays. It can destroy body cells, and kill you. But it is also very useful.

Some medical uses

- In treating cancer, radiation is used to kill cancer cells. Cobalt-60 is usually used for this.
- Radioisotopes are used as medical tracers. For example iodine-131 is used to see if the thyroid gland is working properly. (The thyroid absorbs iodine.) A tiny amount of iodine-131 is injected, and the radiation from it is tracked using a special camera.

Some industrial uses

- Radioisotopes are used as tracers to detect leaks in oil and gas pipes. Radiation can be detected at the leak, using an instrument called a Geiger counter.
- Radiation is also used to **sterilize** things (kill the bacteria in them). In many countries it is used to sterilize foods such as fruit and meat, to keep them fresh for longer.

✓

Quick check for 3.2   *(Answers on page 164)*

1 What are isotopes?
2 Carbon has three isotopes: $^{12}_{6}\text{C}$ $^{13}_{6}\text{C}$ $^{14}_{6}\text{C}$

   a How many neutrons does each isotope have?
   b The average mass for carbon atoms is very close to 12. So which of the three isotopes of carbon is the most common?
   c One of the isotopes is called carbon-14. Which one?
3 Carbon-14 is radioactive. What does that mean?

3.3 More about the electron arrangement

The electrons in an atom are arranged in shells around the nucleus, like this:

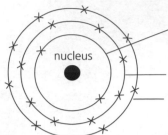

The first shell can hold 2 electrons.
It fills up first. (It is at the lowest energy level.)

The second shell can hold 8 electrons. It fills up next.

The third shell can hold 18 electrons. But it fills up to 8.
Then the next 2 go into the fourth shell (not shown here).
After that, the rest of the third shell fills.

The distribution of the electrons in their shells, in this atom, is shown as: **2 + 8 + 8**.
(Or sometimes as 2, 8, 8 or 2.8.8.)

✔

Quick check for 3.3   *(Answers on page 164)*

1   The drawing on the right shows the arrangement of electrons in an atom.
Which describes this electron distribution correctly?
a 1 + 8 + 2   **b** 2 + 8 + 1   **c** 1 + 7 + 1

2   Draw a sketch like the one on the right for an atom with the electron
distribution 2 + 8 + 6.

3   Draw a sketch like the one on the right for an atom that has:
a 1 electron   **b** 3 electrons   **c** 19 electrons

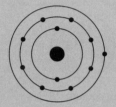

3.4 Protons, electrons, and the Periodic Table

This shows the electron arrangement for the atoms of the first 20 elements in the Periodic
Table. Notice how the shells fill up in order.

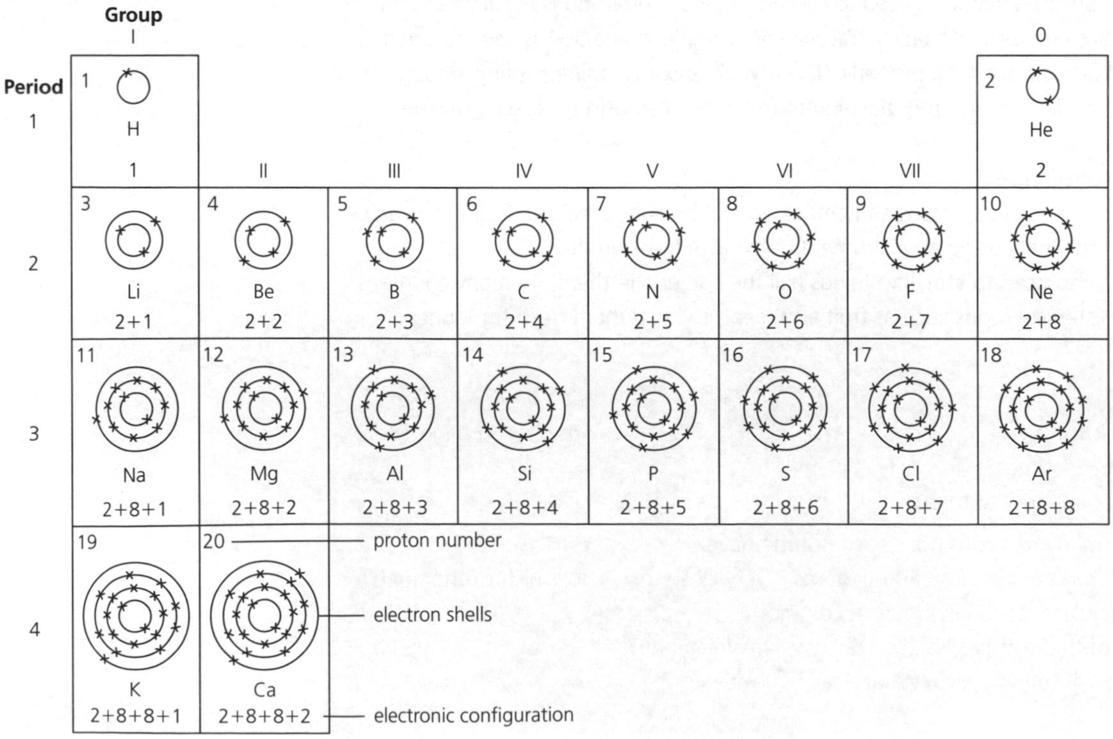

Group I	II	III	IV	V	VI	VII	0
Period 1 — 1 H 1							2 He 2
3 Li 2+1	4 Be 2+2	5 B 2+3	6 C 2+4	7 N 2+5	8 O 2+6	9 F 2+7	10 Ne 2+8
11 Na 2+8+1	12 Mg 2+8+2	13 Al 2+8+3	14 Si 2+8+4	15 P 2+8+5	16 S 2+8+6	17 Cl 2+8+7	18 Ar 2+8+8
19 K 2+8+8+1	20 Ca 2+8+8+2						

proton number

electron shells

electronic configuration

Notes about the Periodic Table
- The elements are arranged in order of their proton number, row by row.
- The rows are called **periods**, and the columns are called **groups**.
- The period number shows how many shells there are.
- The group number for Groups I to VII is the same as the number of outer-shell electrons.
- The outer-shell electrons are called the **valency electrons**. They dictate how the element reacts. So all the elements in Group I have similar reactions, because their atoms all have one valency electron.
- The atoms of the Group 0 elements (the noble gases) have 8 electrons in their outer shells, except for helium, which has 2. This arrangement makes these atoms stable. So the Group 0 elements are unreactive.

helium atom
full outer shell of 2 electrons
stable

neon atom
full outer shell of 8 electrons
stable

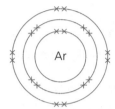
argon atom
outer shell of 8 electrons
stable

- Note how the electrons are shown in pairs, above. As you will see later, when shells are more than half 'full', the electrons begin to pair up.

The elements after calcium
The table on page 18 shows the first 20 elements, up to calcium. After calcium, the shells fill in a more complex order. But you can answer questions about elements after calcium, if you remember the relationship between group number, and valency (outer-shell) electrons. Look at these examples. (You can find the elements in the Periodic Table on page 105.)

Element	Proton number (= number of electrons too)	Where the element is in the Periodic Table		Electron distribution
		Period (tells you how many shells its atoms have)	**Group** (tells you how many valency electrons there are; Group 0 has a full outer shell)	
bromine, Br	35	4	VII	2 + 8 + 18 + 7
krypton, Kr	36	4	0	2 + 8 + 18 + 8
rubidium, Rb	37	5	I	2 + 8 + 18 + 8 + 1
strontium, Sr	38	5	II	2 + 8 + 18 + 8 + 2

Quick check for 3.4   (Answers on page 164)
1 a Name the element made up of atoms with:
     i 3 electrons   ii 9 electrons   iii 17 electrons
   b Which two of the elements in a will show similar chemical properties? Why?
2 Name the element that has 15 protons, and give its electron distribution.
3 Name the element in: a Group 0, Period 1   b Group V, Period 3
4 Name the elements with these electron distributions:
   a 2   b 2 + 5   c 2 + 8 + 2   d 2, 8, 7   e 2.8.8.2
5 Give the missing number for these atoms:
   a proton number 1, electron distribution ___
   b proton number 12, electron distribution ___ + 8 + ___
   c proton number 37, electron distribution 2 + 8 + ___ + ___ + 1

Extended

Questions on section 3

Answers for these questions are on page 164.

Core curriculum

1 Look at the list of five elements below.

argon
bromine
chlorine
iodine
potassium

- **a** Put the five elements in order of increasing proton number.
  (The Periodic Table on page 105 will help you.)
- **b** Put the five elements in order of increasing relative atomic mass.
- **c** The orders in your answers for **a** and **b** are different.
  Which one of these is the most likely explanation for the difference?
  - A   The proton number of a particular element may vary.
  - B   The presence of neutrons.
  - C   The atoms easily gain or lose electrons.
  - D   The number of protons must always equal the number of neutrons.
- **d** Which of the five elements in the list are in the same group of the Periodic Table?
- **e   i** From the list, choose one element which has one electron in its outer shell.
  - **ii** From the list, choose one element which has a full outer shell of electrons.

CIE 0620 June '04 Paper 2 Q5

2 Bromine is an element in Group 7 of the Periodic Table.
- **a** Bromine has two isotopes.
  The nucleon (mass number) of bromine-79 is 79 and of bromine-81 is 81.
  - **i** What is the meaning of the term isotopes?
  - **ii** Copy and complete this table to show the numbers of subatomic particles in atoms of bromine-79 and bromine-81. (See the Periodic Table on page 105.)

Number of	bromine-79	bromine-81
electrons		
neutrons		
protons		

3 Iron has several isotopes.
- **a** The table shows the number of subatomic particles in one isotope of iron.

Type of particle	Number of particles	Relative charge on the particle
electrons	26	
neutrons	30	
protons	26	

Complete the table to show the relative charge on each particle.
- **b** State the number of nucleons in this isotope of iron.
- **c** Some isotopes are radioactive. State one industrial use of radioactive isotopes.

CIE 0620 November '03 Paper 2 Q4

Extended curriculum

1 a Complete this table for the three subatomic particles.

Name	Symbol	Relative mass	Relative charge
electron	e⁻		
proton		1	
	n		0

b Use the information in the table to explain the following.
   i Atoms contain charged particles but they have no overall charge.
   ii Atoms can form positive ions.
   iii Atoms of the same element can have different masses.
   iv Scientists are certain that there are no undiscovered elements missing from the Periodic Table from hydrogen to lawrencium.   *CIE 0620 June '08 Paper 3 Q2*

2 Calcium and argon both form particles of similar mass.
   a Copy and complete the following table that shows the number of protons, electrons and neutrons in each particle:

Particle	Protons	Electrons	Neutrons
^{40}Ar			
^{40}Ca			
^{44}Ca			

   b Explain why: **i** ^{40}Ca and ^{44}Ca are isotopes   **ii** ^{40}Ca and ^{40}Ar are not isotopes.
   c Complete the electron distribution of an atom of calcium:
   $$2 + 8 + \ldots\ldots + \ldots\ldots$$

3 The first three elements in Period 5 of the Periodic Table of the Elements are rubidium, strontium and yttrium.
   a How many **more** protons, electrons and neutrons are there in one atom of yttrium than in one atom of rubidium? The Periodic Table on page 105 will help you.
   number of protons ....
   number of electrons ....
   number of neutrons ....
   b Complete the electron distribution of an atom of strontium:
   $$2 + 8 + 18 + \ldots\ldots + \ldots\ldots$$

4 The table below gives the number of protons, neutrons and electrons in atoms or ions.

Particle	Number of protons	Number of electrons	Number of neutrons	Symbol or formula
A	9	10	10	$^{19}_{9}$F⁻
B	11	11	12	
C	18	18	22	
D	15	18	16	
E	13	10	14	

   a Write the correct symbol or formula for B, C, D and E.
   b Which atom in the table is an isotope of the atom with the composition 11p, 11e and 14n? Give a reason for your choice.   *CIE 0620 November '07 Paper 3 Q2*

4 Bonding and structure

The big picture

- Atoms can join or 'bond' together. The purpose is to gain an outer shell of electrons like the noble gas atoms have, because that is a stable arrangement.
- There are three types of bonding: covalent, ionic, and metallic.
- The bonded atoms form a regular structure or lattice, in the solid state.
- The properties of a substance are influenced by both bonding and structure.

4.1 First, a review of some basic ideas

Elements, compounds and mixtures

Substance	What is in it?	Example
element	contains only one type of atom.	— the element carbon contains only carbon atoms
compound	contains more than one type of atom, held together by chemical bonds: ➤ the atoms are always bonded in the same ratio ➤ you cannot separate them by physical means.	—in the compound carbon dioxide, each carbon atom is always bonded to two oxygen atoms
mixture	can contain any number of different substances, in any ratio: ➤ the substances are just mixed, and *not* joined by chemical bonds. ➤ you can separate them using one of the separation methods in Section 2.	mixture of salt and sand you can mix any amounts of salt and sand, and separate them by adding water to dissolve the salt, then filtering

Atoms and ions

Atoms and ions are closely related. What is the difference between them?

Atoms	Ions	
contain equal numbers of protons and electrons, so have no charge.	are atoms or group of atoms that carry a charge, because they have gained or lost electrons.	
Examples of atoms	**Example of a positive ion**	**Example of a negative ion**
A **sodium atom** has 11 protons and 11 electrons, so its charge is:	A **sodium ion** is a sodium atom that has *lost* an electron, so its charge is:	A **chloride ion** is a chlorine atom that has *gained* an electron, so its charge is:
from the protons 11+ from the electrons 11− total 0	from the protons 11+ from the electrons 10− total 1+	from the protons 17+ from the electrons 18− total 1−
A **chlorine atom** has 17 protons and 17 electrons, so its charge is: from the protons 17+ from the electrons 17− total 0	We show it as Na^+. It is called a positive ion because it has a positive charge.	We show it as Cl^-. It is called a negative ion because it has a negative charge.

Metals and non-metals

There are over 100 elements. They are all different. But they can be divided into two groups: **metals** and **non-metals**. As you will see later, the differences between the two groups are due to differences in bonding and structure.

Metals	Non-metals
readily conduct electricity and heat	do not conduct electricity or heat
are mostly malleable (can be hammered into shape) and ductile (can be drawn into wires)	are usually brittle when solid – they break up when they are hammered
tend to be shiny	look dull, when solid
tend to have high density (they are heavy)	have low density
usually have high melting points	have low melting points (many are gases at room temperature)
form positive ions, in reactions	when they form ions, these are negative (except for H^+)
usually form basic oxides (see page 88)	usually form acidic oxides (see page 88)

Some exceptions

- The metal mercury is a liquid at room temperature.
- Some metals, such as sodium and potassium, have low density and low boiling points.
- Hydrogen is a non-metal, but it forms positive ions, H^+.
- Diamond and graphite are two forms of carbon, a non-metal. Graphite is a soft greasy solid, and a good conductor of electricity. Diamond is very hard, with a very high melting point. Graphite and diamond have special structures, as you will see later.

Metals and alloys

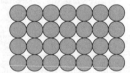

Metals are **elements** so they contain only **one** type of atom.

Alloys are **mixtures**, where at least one other substance is added to the metal.

The added substances in alloys can be other metals, or non-metals such as carbon. They are chosen to improve on certain properties of a metal, and make it more useful.

Examples of alloys

Alloy	What is in it, and typical %	Special properties	Uses
brass	copper 70% zinc 30%	much harder than copper; and unlike copper it does not corrode	musical instruments, ornaments, door knobs and other fittings
stainless steel	iron 70% chromium 20% nickel 10%	unlike iron, it does not corrode	car parts, kitchen sinks, cutlery, tanks in chemical factories

Quick check for 4.1 (Answers start on page 164)
1 Why is iron sulfide (FeS) classed as a compound?
2 Give three ways in which iron sulfide differs from a mixture of iron and sulfur.
3 Iron is a metal, and sulfur is a non-metal. Give three ways in which you expect them to behave differently.
4 Iron rusts (corrodes) easily. How would you improve on its properties?

4.2 An overview of bonding and structure

- Bonding is about how atoms are joined together.
- Structure is about how the bonded atoms are arranged.

Bonding
There are three types of bonding: covalent, ionic and metallic.

Type of bonding	covalent	ionic	metallic
Description	electrons are shared between atoms	electrons are transferred from one atom to another, forming ions	a lattice of positive ions in a sea of electrons
What kinds of atoms bond together in this way?	non-metal atoms: ➤ of the same element, or ➤ of different elements, giving compounds	metal atoms bond with non-metal atoms, to give compounds	➤ only metal atoms ➤ they are usually atoms of the same metal ➤ but an alloy has atoms of different metals
What holds the atoms together?	the bonds created by sharing electrons	the attraction between ions of opposite charge	the attraction between the positive ions and the electrons

Why do atoms bond?
Atoms bond to other atoms in order to gain the same arrangement of outer-shell electrons as a noble gas atom – because that is a stable arrangement. (See page 19.)

Remember

Outer shells of noble gas atoms:
helium – 2 electrons
neon – 8 electrons
argon – 8 electrons

Structure
In the solid state, the particles form a regular arrangement called a **lattice**.
There are two types of structure:

Structure	simple molecular	giant
Description	the lattice is built up of millions of separate small molecules	the lattice is built up of millions of particles, which can be: ➤ positive and negative ions, joined by ionic bonds, or ➤ metal ions in a sea of electrons, joined by metallic bonds, or ➤ non-metal atoms joined by covalent bonds
Example	in iodine (I_2) the lattice is made up of iodine molecules, each containing two atoms	in sodium chloride (NaCl) the lattice is made up of sodium and chloride ions
What holds the structure together?	strong covalent bonds within molecules, but weak forces between molecules	particles held together by a network of strong bonds
One result of this structure	the solid has a low melting point, since it does not take much heat energy to break up the lattice to form a liquid	the solid usually has a high melting point, since it takes a great deal of heat energy to break the bonds in the lattice

✓ **Quick check for 4.2**   *(Answers on page 165)*
1 Explain the difference between *bonding* and *structure*.
2 What is the main difference between an ionic bond and a covalent bond?
3 What is a *lattice*?
4 Give another name for a small group of atoms joined by covalent bonds.
5 Using information given above, describe the bonding in silver.

Extended

4.3 Ionic bonding

- Ionic bonds form only between metal and non-metal atoms.
- Electrons are transferred from the metal atom to the non-metal atom, to give ions of opposite charge.
- The ions are stable, because they have the same arrangement of outer-shell electrons as a noble gas atom does.
- The ions are then attracted to each other. (Opposite charges attract.)

Examples of ionic bonding

Sodium chloride

A sodium atom transfers its outer electron to a chlorine atom, giving a positive sodium ion and a negative chloride ion. These ions are stable because they have 8 electrons in their outer shells, like neon and argon atoms:

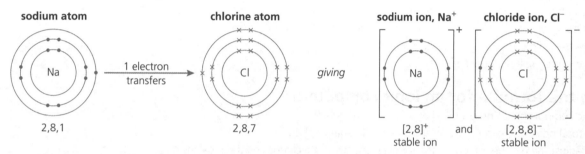

The bonding is the same in *any* compound formed between an alkali metal (Group I) and a halogen (Group VII). For example in lithium bromide (LiBr), and potassium iodide (KI).

The structure of sodium chloride

Millions of ions group together to form a lattice. The lattice is a regular arrangement of alternating positive and negative ions. They are held together by the strong ionic bonds between ions of opposite charge:

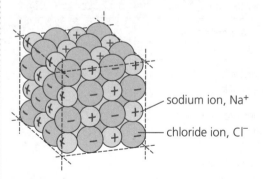

sodium ion, Na^+

chloride ion, Cl^-

Magnesium oxide

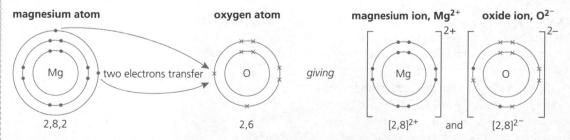

Bonding: the attraction between positive magnesium ions and negative oxide ions.
Structure: a lattice containing equal numbers of magnesium and oxide ions.

Extended

Magnesium chloride

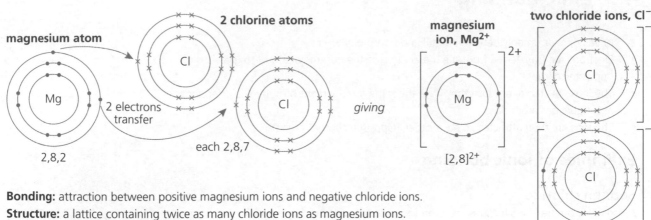

magnesium atom

2 chlorine atoms

Mg

2,8,2

2 electrons transfer

Cl

Cl

each 2,8,7

giving

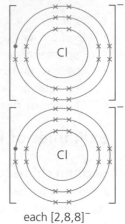

magnesium ion, Mg²⁺

Mg

$[2,8]^{2+}$

two chloride ions, Cl⁻

Cl

Cl

each [2,8,8]⁻

Bonding: attraction between positive magnesium ions and negative chloride ions.
Structure: a lattice containing twice as many chloride ions as magnesium ions.

The properties of ionic compounds

- They have high melting and boiling points, because the bonds between ions are strong.
- They conduct electricity when melted or dissolved in water, because the charged ions are then free to move.
- They are usually soluble in water.

Writing the formula for an ionic compound

- An ionic compound has no overall charge.
- So the total positive charge must balance the total negative charge.
- So you adjust the number of ions in the formula, until the total charges on them balance.

Examples

Ionic compound	positive ion	negative ion	balancing their charges	so the formula is
magnesium oxide	Mg^{2+}	O^{2-}	they are balanced (2⁺ and 2⁻)	MgO
magnesium chloride	Mg^{2+}	Cl^-	two Cl⁻ ions are needed to balance one Mg²⁺ ion	$MgCl_2$
sodium oxide	Na^+	O^{2-}	two Na⁺ ions are needed to balance one O²⁻ ion	Na_2O
aluminium hydroxide	Al^{3+}	OH^-	three OH⁻ ions are needed to balance one Al³⁺ ion	$Al(OH)_3$
calcium nitrate	Ca^{2+}	NO_3^-	two NO₃⁻ ions are needed to balance one Ca²⁺ ion	$Ca(NO_3)_2$

Look at the last two formulae above. They show how to use brackets, if there is more than one unit of the compound ion. (A compound ion contains atoms of different elements.)

✓

Quick check for 4.3   *(Answers on page 165)*

1 Explain *how* a sodium atom becomes a sodium ion, and *why*.
2 Draw a diagram to show the electron transfer when a lithium atom reacts with a fluorine atom.
3 Describe the structure of a solid ionic compound.
4 Draw a diagram to show the ionic bond formed between:
   a calcium and oxygen atoms   **b** calcium and chlorine atoms
5 Write the formula for: **a** sodium hydroxide   **b** aluminium chloride
   c magnesium hydroxide

4.4 Covalent bonding: simple molecules

Covalent bonds are formed when atoms share electrons.
- They are formed:
  - ➤ between the atoms in a non-metal element
  - ➤ between atoms of different non-metals, to give a compound.
- The purpose is to gain the same arrangement of outer-shell electrons as a noble gas atom – because that is a stable arrangement.
- The bonded atoms form a unit called a **molecule**.

> ### Remember
> Outer-shell electrons in the noble gases:
> helium – 2 electrons (full shell)
> neon – 8 electrons (full shell)
> argon – 8 electrons

The molecules can be simple molecules with a small number of atoms, or giant molecules (macromolecules), with millions of atoms. In this section we concentrate on the simple molecules. Note that in the drawings below, only the outer electron shells are shown.

Molecular elements
Let's look first at the bonding in some non-metal elements.

Hydrogen
A hydrogen atom has 1 electron, but needs 2 for a full shell, like a helium atom has.
So each shares its electron with another hydrogen atom, to form a hydrogen molecule, H_2.
The covalent bond is the attraction between the positive nuclei and the shared electrons.

two hydrogen atoms   **a hydrogen molecule, H_2**

a shared pair of electrons

Chlorine
A chlorine atom shares an outer electron with another chlorine atom, to form a chlorine molecule, Cl_2. Each atom obtains an outer shell of 8 electrons, like an argon atom has.

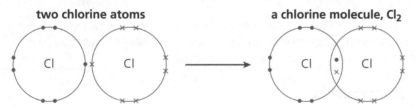

two chlorine atoms   **a chlorine molecule, Cl_2**

Oxygen
An oxygen atom shares two outer electrons with another oxygen atom, to form an oxygen molecule, O_2. Each atom obtains a full outer shell of electrons (8), like a neon atom has. Since two atoms are shared, the bond is called a **double** covalent bond (O=O).

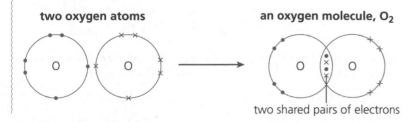

two oxygen atoms   **an oxygen molecule, O_2**

two shared pairs of electrons

Extended

Extended

Nitrogen

A nitrogen atom shares three outer electrons with another nitrogen atom to form a nitrogen molecule, N_2. Each atom obtains a full outer shell of electrons (8), like a neon atom has. Since three atoms are shared, the bond is called a **triple** covalent bond (N≡N).

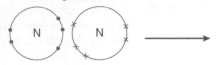

two nitrogen atoms

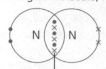

a nitrogen molecule, N_2

three shared pairs of electrons

> **Remember**
> If an atom has 4 electrons in its outer shell, they are not paired. But each extra electron after that pairs up. (Look at a nitrogen atom.)

Molecular compounds

In a molecular compound, atoms share outer electrons with different atoms to form molecules. Each atom obtains an outer shell of electrons like a noble gas atom has. (Hydrogen obtains 2, others 8.) Look at these examples:

Water, H_2O	Methane, CH_4	Hydrogen chloride, HCl

Their shapes

Pairs of electrons repel each other, so they try to get as far apart as possible. That influences the shape of the molecule. Here are models of the three molecules above:

Water	Methane	Hydrogen chloride
The atoms are not in a straight line because all the electron pairs (shared and unshared) repel each other.	The four pairs of shared electrons have moved as far apart as possible, giving a tetrahedral shape.	Here there are only two atoms, so only one possible shape: they lie in a straight line.

More examples of molecular compounds

Ethene, C_2H_4	Carbon dioxide, CO_2	Methanol, CH_3OH
The molecule has a mix of single and double bonds. You can show it as:	The molecule has two double bonds. You can show it as:	The molecule has three different types of atom. You can show it as:

Extended

Structure of simple molecular substances

In the solid, the molecules are arranged in a lattice. Look at the forces holding it together:

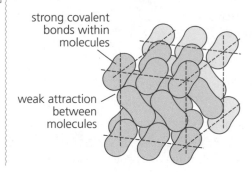

strong covalent
bonds within
molecules

weak attraction
between
molecules

Properties of simple molecular substances

Substances with a simple molecular structure:

*   have low melting and boiling points. This is because the intermolecular forces are weak. It does not take much heat energy to break up the lattice, and separate the particles from each other. (That is why many molecular substances are gases at room temperature.)
*   do not conduct electricity, because they have no charge.
*   are usually insoluble in water, but soluble in organic solvents (for example propanone).

Comparing ionic and molecular compounds

Molecular compounds (such as methane, CH_4)	Ionic compounds (such as sodium chloride, NaCl)
• have low melting and boiling points; many are gases or liquids at room temperature	• have high melting and boiling points, so they are solids at room temperature
• evaporate readily – they are volatile	• are not volatile
• do not conduct electricity	• conduct electricity
• are insoluble in water, but dissolve in organic solvents	• are usually soluble in water

✓
Quick check for 4.4   (*Answers on page 165*)
1   Explain how two hydrogen atoms become a hydrogen molecule.
2   Why are molecules stable?
3   Draw a diagram to show what happens to the electrons when carbon reacts with chlorine to form tetrachloromethane, CCl_4. (Show outer-shell electrons only.)
4   If you cool hydrogen gas down enough, it will become a liquid, and then freeze to a solid. Describe the structure of this solid.
5   Molecular compounds evaporate easily – they are *volatile*. Why is this?
6   Draw a diagram to show the covalent bonding in:
    a carbon dioxide   **b** nitrogen
7   Ammonia (NH_3) is a covalent compound. Draw a diagram to show the bonding in an ammonia molecule. (Show outer electrons only.)

4.5 Covalent bonding: macromolecules

Some substances with covalent bonding form **macromolecules**. A macromolecule is
a giant structure (or lattice) of millions of atoms, all held together by covalent bonds.

Carbon: a macromolecular element

Carbon is a non-metal. It occurs in two forms or **allotropes**: diamond and graphite.
Both are macromolecular. Compare them:

Form	diamond	graphite
Bonding	A carbon atom shares **all four** of its outer electrons with other carbon atoms, to form a three-dimensional lattice.	A carbon atom shares **three** of its outer electrons with other carbon atoms, to form a layer structure. The fourth electron exists between the layers and is free to move (like electrons in metals – see page 31).
Giant structure	strong covalent bonds Each carbon atom forms a **tetrahedron** with four other carbon atoms.	strong covalent bonds weak forces Each carbon atom becomes part of a **flat hexagonal ring**.
Forces	All the covalent bonds are identical, and strong. There are no weak forces.	The covalent bonds **within** the layers are strong. But the layers are held together by weak forces.
Properties	• very high melting point, because all the bonds are strong • very hard, for the same reason • non-conductor of electricity, because there are no electrons free to move • insoluble in water	• very high melting point, because the covalent bonds are strong • soft and slippery, because the layers slide over each other easily • good conductor of electricity, because the 'free' electrons between the layers can move • insoluble in water
Uses	cutting tools jewellery	lubricant for engines and locks electrodes for electrolysis in the lab and in industry

Silicon dioxide: a macromolecular compound

Silicon dioxide, SiO_2, is a macromolecular compound. It occurs naturally as sand and quartz.

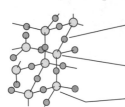 each oxygen atom ● forms covalent bonds with two silicon atoms ○

each silicon atom forms covalent bonds with four oxygen atoms

the silicon and oxygen atoms form a tetrahedron (like the carbon atoms in diamond)

- The bonds are strong, as in diamond. So silicon dioxide has similar properties to diamond:
  ➤ it has a very high melting point, and is very hard
  ➤ it is a non-conductor of electricity, and insoluble in water.
- It is used in sandpaper, and to line furnaces. (Since it occurs widely in nature, it is cheap.)

Extended

✓
Quick check for 4.5   *(Answers on page 165)*
1   What holds the atoms together, in a macromolecule?
2   In which ways are molecules and macromolecules:
     a the same?   **b** different?
3   Graphite is soft and slippery, and a good conductor of electricity. Explain why it
     has these properties.
4   Silicon dioxide has similar properties to diamond – but not to graphite.
     Explain why.

4.6 Metallic bonding

Extended

*   Metallic bonds are the bonds between metal atoms, in a metal or metal alloy.
*   The outer electrons leave the metal atoms, giving metal ions with full outer shells.
    For example, in silver:

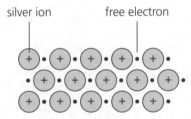

silver ion   free electron

*   The outer electrons form a sea of free electrons around the metal ions, in a giant lattice.
    (They are called 'free' electrons because they can move around freely.)
*   Metallic bonds are the result of the attraction between the positive metal ions, and the
    free electrons.
*   Metallic bonds are strong.

Explaining the properties of metals

This table shows how the bonding and structure in metals account for some of their
properties:

Properties of metals	Reasons
They usually have high melting and boiling points.	They form a giant lattice, with strong bonds.
They conduct electricity, when solid and melted.	The 'free' electrons carry the electric charge through the metal.
They are malleable: they can be readily bent, pressed, or hammered into shape.	The layers of atoms can slide over each other, while the free electrons can also move (so the metallic bond is not broken).
They are ductile: they can be drawn into wires.	Same reason as above.

✓
Quick check for 4.6   *(Answers on page 165)*
1   Describe the metallic bond.
2   What type of structure does a metal have?
3   All metals are good conductors of electricity. Why?

Questions on section 4

Answers for these questions are on page 165.

Core curriculum

1 a Complete the diagram to show the electronic structure of water:
show **hydrogen** electrons by o
show **oxygen** electrons by ×

H   H

b The structure of phosphorus trioxide is shown below.

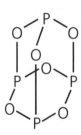

Write the simplest formula for phosphorus trioxide.

CIE 0620 November '07 Paper 2 Q1

2 The structures of some elements and compounds are shown below.

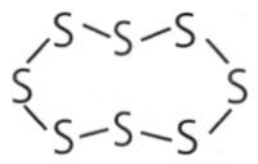

A 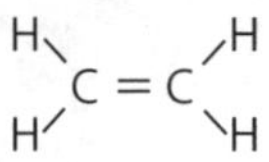   B 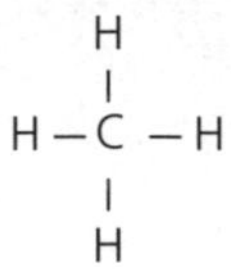   C

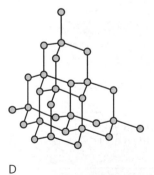 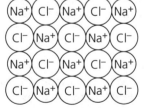

D   E 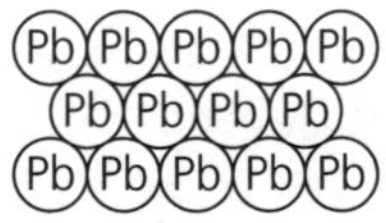   F

Answer these questions using the letters A to F.

i   Which structure is methane?
ii Which structure contains ions?
iii Which structure is a metal?
iv Which structure is sodium chloride?
v   Which structure is diamond?
vi Which structure contains a double covalent bond?
vii Which **three** structures are elements?

CIE 0620 June '07 Paper 2 Q1

3 Carbon exists in two forms, graphite and diamond.

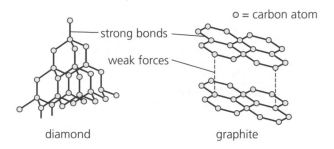

o = carbon atom

strong bonds

weak forces

diamond   graphite

Use ideas about structure and bonding to suggest:
a why graphite is used as a lubricant,
b why diamond is very hard.   *CIE 0620 November '07 Paper 3 Q6*

Extended curriculum
1 The structural formula of carbonyl chloride is given below.

$$O=C\begin{smallmatrix}/Cl\\\backslash Cl\end{smallmatrix}$$

Draw a diagram that shows the arrangement of the valency electrons in one molecule of this covalent compound.
Use x for an electron from a chlorine atom.
Use o for an electron from a carbon atom.
Use • for an electron from an oxygen atom.   *CIE 0620 June '08 Paper 3 Q2*

2 Complete the following table.

Type of structure	Particles present	Electrical conductivity of solid	Electrical conductivity of liquid	Example
ionic	positive and negative ions	poor	**i**..........................	**ii**..........................
macromolecular	atoms of two different elements in a giant covalent structure	poor	poor	**iii**..........................
metallic	**iv**.......................... and **v**..........................	good	**vi**..........................	copper

CIE 0620 June '07 Paper 3 Q2

3 Magnesium reacts with bromine to form magnesium bromide.
  a Magnesium bromide is an ionic compound. Draw a diagram that shows the formula of the compound, the charges on the ions, and the arrangement of outer electrons around the negative ion.
    The electron distribution of a bromine atom is 2, 8, 18, 7.
    Use x to represent an electron from a magnesium atom.
    Use o to represent an electron from a bromine atom.
  b In the lattice of magnesium bromide, the ratio of magnesium ions to bromide ions is 1:2.
    i Explain the term *lattice*.
    ii Explain why the ratio of the ions is 1:2.   *CIE 0620 November '07 Paper 3 Q3*

5 Ratios and amounts, in reactions

The big picture

- Chemistry is all about substances reacting with each other.
- Chemical equations are just a short and efficient way to describe how substances react.
- From a chemical equation, you can work out:
  - ➤ what the reactants and products are
  - ➤ the ratios of the substances taking part in the reaction
  - ➤ amounts of substances, in grams (or kilograms, or tonnes).
- The study of the ratios in which substances react is called stoichiometry.

5.1 Chemical formulae and equations

Chemical formulae

- Atoms always combine in a fixed ratio.
- They combine in the ratio that will give them full outer shells of electrons.
- You can tell what the ratio is, from the chemical formula.

Example	water	carbon dioxide
Formula	H_2O	CO_2
Ratio in which atoms combine	**two** atoms of hydrogen combine with **one** atom of oxygen	**two** atoms of oxygen combine with **one** atom of carbon

Chemical equations

- **Chemical equations** are a short way to describe a reaction.
- You can give them in words, as **word equations**. Or you can use formulae, giving **symbol equations**, which provide much more information.
- The + sign in an equation means *and*. The arrow means *react(s) to give*.

Example	burning of hydrogen in oxygen	burning of carbon in oxygen
Word equation	hydrogen + oxygen $\longrightarrow$ water	carbon + oxygen $\longrightarrow$ carbon dioxide
Balanced symbol equation	$2H_2\,(g) + O_2\,(g) \longrightarrow 2H_2O\,(l)$	$C\,(s) + O_2\,(g) \longrightarrow CO_2\,(g)$
Ratios of atoms and molecules in the reaction	**two** molecules of hydrogen and **one** molecule of oxygen react to give **two** molecules of water	**one** atom of carbon and **one** molecule of oxygen react to give **one** molecule of carbon dioxide

✓
Quick check for 5.1   *(Answers on page 165)*

1. In these compounds, which atoms are combined, and in what ratio?
   - **a** methane, CH_4
   - **b** ammonia, NH_3
   - **c** iron sulfide, FeS
   - **d** calcium carbonate, $CaCO_3$.
2. Write the word equation for this reaction:
   $CH_4\,(g) + 2O_2\,(g) \longrightarrow CO_2\,(g) + 2H_2O\,(l)$
3. Give the ratios for all the molecules that take part in the reaction in question **2**.

5.2 Writing balanced chemical equations

This table shows how to write a balanced chemical (symbol) equation.

Steps	Worked example
1 Write the word equation.	calcium + water $\longrightarrow$ calcium hydroxide + hydrogen
2 Write the correct formulae.	$Ca + H_2O \longrightarrow Ca(OH)_2 + H_2$
3 Count atoms on each side of the arrow. Are the numbers equal?	Ca: **1** on left, **1** on right H: **2** on left, **4** on right O: **1** on left, **2** on right The numbers are not equal, so the equation is not balanced.
4 Try putting numbers *in front of* formulae, to balance the atoms. (You must not change any formulae.)	There is twice as much hydrogen on the right, so try a 2 in front of H_2O on the left: $Ca + 2H_2O \longrightarrow Ca(OH)_2 + H_2$ Check atoms again … Ca: **1** on left, **1** on right H: **4** on left, **4** on right O: **2** on left, **2** on right so the equation is now balanced.
5 Add **state symbols** to show the state of the substances taking part in the reaction: (*s*) for solid (*l*) for liquid (*g*) for gas (*aq*) for aqueous solution (a solution in water).	$Ca\,(s) + 2H_2O\,(l) \longrightarrow Ca(OH)_2\,(aq) + H_2\,(g)$

✓

Quick check for 5.2   (*Answers on page 165*)
1 Write a balanced symbol equation for this reaction: 1 molecule of hydrogen and 1 molecule of chlorine react to give 2 molecules of hydrogen chloride.
2 The formulae below are correct. But are the equations balanced? If not, balance them. Then see if you can add the correct state symbols:
   a $Mg + O_2 \longrightarrow MgO$ **b** $Fe + O_2 \longrightarrow Fe_2O_3$
3 The colourless liquid hydrogen peroxide (H_2O_2) decomposes slowly to water and oxygen. Write a balanced symbol equation for the reaction, with state symbols.

5.3 Relative atomic and molecular mass

Relative atomic mass, A_r
An atom is so tiny that we can't measure its mass in grams. Instead, the carbon-12 atom is taken as the standard, and assigned a mass of 12 units. Other atoms are compared to it:

It has: 6 protons 6 neutrons C-12	Mg	H
The mass of the C-12 atom is taken as 12. (That fits with a mass of 1 for each proton and neutron.)	This magnesium atom was found to have twice the mass of a carbon-12 atom. So its mass is 24.	This hydrogen atom was found to have 1/12 the mass of a carbon-12 atom. So its mass is 1.

The mass of an atom, compared to carbon-12, is called its **relative atomic mass**.
It is written in a short way as A_r.

A_r values for some common elements

Element	Symbol	A_r
carbon	C	12
hydrogen	H	1
nitrogen	N	14
oxygen	O	16
sodium	Na	23
sulfur	S	32
chlorine	Cl	35.5

Why does chlorine has an A_r of 35.5?
Chlorine has two **isotopes** (page 16).
75% of its atoms are chlorine-35, and 25% are chlorine-37.

So A_r for chlorine $= \frac{75}{100} \times 35 + \frac{25}{100} \times 37 = \mathbf{35.5}$

Since 35.5 is halfway between 35 and 36, this A_r is not rounded off for calculations.

Relative molecular mass (M_r)

- The mass of a molecule is called its **relative molecular mass** or M_r.
- You work it out by adding up the A_r values.
- So the M_r for water, H_2O, is: $1 + 1 + 16 = 18$.

Relative formula mass (M_r)

- Many compounds are made of ions, not molecules. We still use the symbol M_r for them, but we call it **relative formula mass**.
- Again you work it out by adding up A_r values. (Ions have the same A_r as atoms, since electrons gained or lost make no difference to the mass.)
- So the M_r for NaCl is: $23 + 35.5 = 58.5$.

More examples of M_r values

Element or compound	Formula	Sum of A_r	M_r	
oxygen	O_2	2×16	32	elements that exist as molecules
nitrogen	N_2	2×14	28	
hydrogen peroxide	H_2O_2	$(2 \times 1) + (2 \times 16)$	34	
carbon monoxide	CO	$12 + 16$	28	
carbon dioxide	CO_2	$12 + (2 \times 16)$	44	molecular compounds
sulfur dioxide	SO_2	$32 + (2 \times 16)$	64	
sodium hydroxide	NaOH	$23 + 16 + 1$	40	
ammonium sulfate	$(NH_4)_2SO_4$	$2 \times [14 + (4 \times 1)] + 32 + (4 \times 16)$	132	ionic compounds

✓

Quick check for 5.3   (Answers on page 165)

1   Define relative atomic mass.
2   An element has atoms that are 4 times heavier than carbon-12 atoms.
    a What is A_r for this element? b Identify it. (Try the Periodic Table on page 105.)
3   A_r for a calcium atom is 40. What is A_r for a calcium ion (Ca^{2+})?
4   Why is the relative atomic mass of chlorine not a whole number?
5   Work out M_r for these compounds:
    a ammonia, NH_3 b methanol, CH_3OH c sodium carbonate, Na_2CO_3
6   For sodium carbonate, M_r stands for *relative formula mass*. What does that tell you about sodium carbonate?

Remember
In working out mass for a compound like $CuSO_4.5H_2O$, $5H_2O$ means 10H and 5O.

5.4 Calculating masses from equations

You can now calculate the masses from chemical equations, using A_r and M_r values.
You just give A_r and M_r as grams. Look at these examples.

Example 1

Balanced equation	$2H_2\,(g) + O_2\,(g) \longrightarrow 2H_2O\,(l)$
Ratios	**two** molecules of hydrogen and **one** molecule of oxygen react to give **two** molecules of water
Using A_r and M_r	$2 \times H_2 = \mathbf{4}$; $O_2 = \mathbf{32}$; $2 \times H_2O = \mathbf{36}$
Giving A_r and M_r as grams	**4 g** of hydrogen and **32 g** of oxygen react to give **36 g** of water
Which means ...	**2 g** of hydrogen and **16 g** of oxygen react to give **18 g** of water
and ...	**1 g** of hydrogen and **8 g** of oxygen react to give **9 g** of water and so on.

Example 2

Balanced equation	$C\,(s) + O_2\,(g) \longrightarrow CO_2\,(g)$
Ratios	**one** atom of carbon and **one** molecule of oxygen react to give **one** molecule of carbon dioxide
Using A_r and M_r	$C = \mathbf{12}$; $O_2 = \mathbf{32}$; $CO_2 = \mathbf{44}$
Giving A_r and M_r as grams	**12 g** of carbon and **32 g** of oxygen react to give **44 g** of carbon dioxide
Which means ...	**6 g** of carbon and **16 g** of oxygen react to give **22 g** of carbon dioxide
and ...	**24 g** of carbon and **64 g** of oxygen react to give **88 g** of carbon dioxide and so on.

So the substances taking part in the reaction are always in the same proportion.

And note that the mass on each side of the arrow is the same.
Mass before reaction = mass after reaction.
The mass does not change, in a chemical reaction.

✓ **Quick check for 5.4**　　　　　　　　　　　　　　*(Answers on page 165)*

1　100 g of reactants react completely, to give a product. What is the mass of this product?

2　a How many grams of oxygen will react with 20 g of hydrogen, to form water?
　　b How many grams of water will form?
　　c 50 g of hydrogen reacted, and 450 g of water was formed. What mass of oxygen was used up?

3　This is the symbol equation for a reaction:
　　$CuCO_3\,(s) \longrightarrow CuO\,(s) + CO_2\,(g)$
　　a Give the word equation for the reaction.
　　b Calculate how much of each product will form, starting with 31 g of the reactant. The A_r values are: Cu, 64; C, 12; O, 16.

Questions on section 5

Answers for these questions are on page 165.
For questions on the Extended curriculum, see Section 6: The mole.

Core curriculum

1   Write a chemical equation for each of these. You do not need to add state symbols.
    Example:

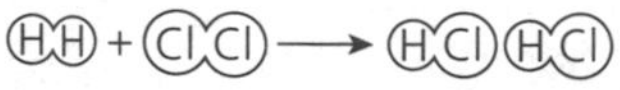

   (H)(H) + (Cl)(Cl) ⟶ (H)(Cl) (H)(Cl)

   $H_2 + Cl_2 \longrightarrow 2HCl$

   a (H)(H) (H)(H) + (O)(O) ⟶ (H)(O)(H) (H)(O)(H)

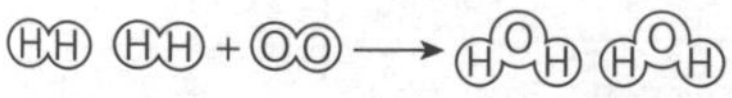

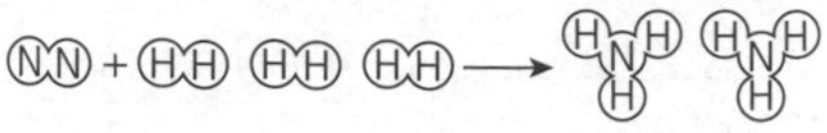

   b (N)(N) + (H)(H) (H)(H) (H)(H) ⟶ (H)(N)(H)(H) (H)(N)(H)(H)

   c (I)(I) + (Cl)(Cl) ⟶ (I)(Cl) (I)(Cl)

   d (P)(P) + (Cl)(Cl) (Cl)(Cl) ⟶ (Cl)(P)(Cl)(Cl) (Cl)(P)(Cl)(Cl)
      (Cl)(Cl)

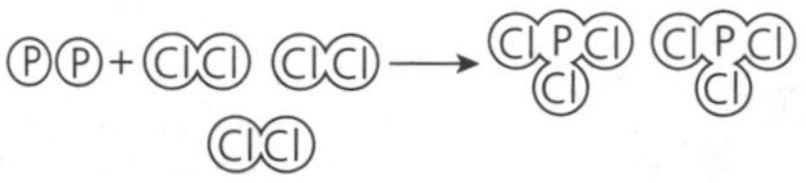

2   Magnesium sulfate can be made by reacting magnesium with dilute sulfuric acid.
    In an experiment a student obtains 24 g of magnesium sulfate from 4.8 g of
    magnesium.
    a   How much magnesium sulfate can the student obtain from 1.2 g of magnesium?
    b   A sample of 20 g of impure magnesium sulfate contains 19.5 g of magnesium
        sulfate. Calculate the percentage purity of the magnesium sulfate.

    CIE 0620 June '08 Paper 2 Q5

3   A small amount of xenon is present in air. Xenon is very unreactive, but a few
    compounds of it have been made in recent years.
    a   Calculate the relative molecular mass of xenon difluoride, XeF_2.
        (A_r: Xe = 131, F = 19)
    b   The structure of another compound of xenon is shown on the right. 
        i   Write the simplest formula for this compound of xenon.
        ii   Calculate the relative molecular mass of this compound. (A_r: O = 16)
        iii Describe the type of bonding in this compound.

    CIE 0620 June '07 Paper 2 Q2

4   Magnesium oxide (MgO) is obtained when magnesium burns in oxygen (O_2).
    a   Construct a balanced chemical equation for the reaction.
    b   Calculate the relative molecular mass of magnesium oxide. (A_r: Mg = 24, O = 16)
    c   2.4 grams of magnesium are burnt in an excess of oxygen.
        i   What mass of magnesium oxide would be formed?
        ii   What mass of oxygen would have been used up?

5   Mercury(II) oxide breaks down into mercury and oxygen when heated:
    $2HgO\ (s) \rightarrow 2Hg\ (l) + O_2\ (g)$
    a   What is the formula mass of mercury(II) oxide? (A_r: Hg = 201, O = 16)
    b   How much mercury and oxygen form when 21.7 g of mercury(II) oxide is heated?

6 The mole

The big picture

- The M_r of hydrogen, H_2, is 2. The M_r of oxygen, O_2, is 32.
- The mole concept tells us that if we could weigh out exactly 2 g of oxygen, and 32 g of hydrogen, they would both contain the same number of molecules.
- This idea helps us to carry out a wide range of calculations, in chemistry.

6.1 Moles and masses

The mole concept

- The mass of an atom of carbon-12 is taken as 12. So the A_r of carbon is 12.
- A magnesium atom has twice this mass. So the A_r of magnesium is 24.
- It follows that 24 grams of magnesium contains the same number of atoms as 12 grams of carbon does.

24 g of magnesium is called **a mole** of magnesium atoms.
A mole of a substance is the amount that contains as many elementary units as the number of atoms in 12 g of carbon-12.
In fact we know how many atoms this is. It is a huge number: **6.02×10^{23}**.
So 24 g of magnesium contains 6.02×10^{23} magnesium atoms.
6.02×10^{23} is called **the Avogadro constant** after the scientist who deduced it.

12 g of carbon powder contain 1 mole of carbon atoms (6.02×10^{23} atoms).

We can apply the same logic to any element, and any compound. Look at these examples:

Substance	A_r or M_r	So 1 mole of it is …	… and contains …
helium, He	4	4 grams	6.02×10^{23} helium atoms
oxygen, O_2	32	32 grams	6.02×10^{23} oxygen molecules

So a mole of a substance is its A_r or M given in grams.

Sample calculations

mass of a given number of moles (g) = mass of 1 mole × number of moles

1	What is the mass of 6 moles of helium atoms?	$4 \times 6 = \textbf{24 g}$
2	What is the mass of 0.5 moles of oxygen molecules?	$32 \times 0.5 = \textbf{16 g}$

$$\text{number of moles in a given mass} = \frac{\text{mass (g)}}{\text{mass of 1 mole}}$$

1	How many moles of helium atoms are in 12 g of helium?	$12 \div 4 = \textbf{3 moles}$
2	How many moles of oxygen molecules are in 80 g of oxygen?	$80 \div 32 = \textbf{2.5 moles}$

Use the calculation triangle!
Cover an item in the triangle below to see how to calculate it. For example: no of moles = mass ÷ mass of 1 mole

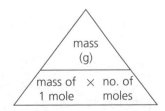

✓
Quick check for 6.1   (Answers on page 165)
1 How many molecules are there in 1 mole of molecules?
2 Find the mass of:
   a 1 mole of chlorine, Cl_2 **b** 0.6 moles of sodium chloride, NaCl
3 Find how many moles of molecules there are in: **a** 8 g of hydrogen, H_2
   b 2.55 g of hydrogen sulfide, H_2S **c** 3.8 g of magnesium chloride, $MgCl_2$

A_r values
H = 1
Na = 23
Mg = 24
S = 32
Cl = 35.5

Extended

6.2 Finding empirical and molecular formulae

- **The empirical formula** shows the simplest ratio in which the atoms in a compound are combined. For example the empirical formula of hydrogen peroxide is HO.
- **The molecular formula** shows the actual number of atoms that combine to form a molecule. The molecular formula of hydrogen peroxide is H_2O_2.

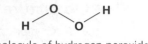

a molecule of hydrogen peroxide

Finding the empirical formula

First, do an experiment to find out what masses of the elements combine to form the compound. Then the empirical formula can be worked out. The steps are:

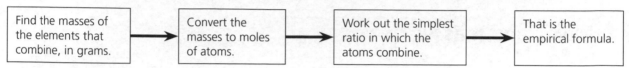

| Find the masses of the elements that combine, in grams. | Convert the masses to moles of atoms. | Work out the simplest ratio in which the atoms combine. | That is the empirical formula. |

Example In an experiment, 80 g of carbon combine with 20 g of hydrogen to form a compound. What is its empirical formula?

Elements that combine	carbon	hydrogen
Masses that combine	80 g	20 g
Relative atomic masses (A_r)	12	1
Moles of atoms that combine	$\frac{80}{12} = 6.67$	$\frac{20}{1} = 20$
Ratio in which atoms combine	6.67 : 20, or 1:3 in its simplest form	

So the empirical formula of the compound is CH_3.

Finding the molecular formula from the empirical formula

The actual formula mass of a compound, M_r, can be found using a machine called a mass spectrometer. From the formula mass and empirical formula, you can work out the molecular formula. The steps are:

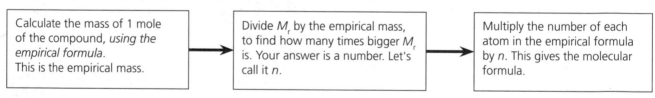

| Calculate the mass of 1 mole of the compound, *using the empirical formula*. This is the empirical mass. | Divide M_r by the empirical mass, to find how many times bigger M_r is. Your answer is a number. Let's call it *n*. | Multiply the number of each atom in the empirical formula by *n*. This gives the molecular formula. |

Example The formula mass of a compound was found to be 30. Its empirical formula was found to be CH_3. What is the molecular formula for the compound?

Empirical mass for the formula CH_3 = 15

$$\frac{M_r}{\text{empirical mass}} = \frac{30}{15} = 2$$

So multiply each atom in the empirical formula CH_3 by 2.
The molecular formula is **C_2H_6**.   (Note: you do *not* put the 2 *in front of* the formula.)

✓

Quick check for 6.2   (*Answers on page 165*)

1   1.2 g of carbon combined with 0.1 g of hydrogen, to form a compound.
   a   Find the empirical formula for the compound.
   b   Its M_r value was found to be 78. Find its molecular formula.
2   In one oxide of nitrogen, the ratio of oxygen atoms to nitrogen atoms is 2:1.
   Its M_r is 46. Give the empirical and molecular formulae for this compound.

A_r values
H = 1
C = 12
N = 14
O = 16

Extended

6.3 Finding masses from chemical equations

Moles and equations

The mole concept lets us work out masses that react. For example:

Ratio of particles in the equation		$2H_2(g)$ 2 molecules	$+$	$O_2(g)$ 1 molecule	$\rightarrow$	$2H_2O(l)$ 2 molecules	
Scaling up to moles gives the mole ratio		2 moles of molecules		1 mole of molecules		2 moles of molecules	
	or	1 mole of molecules		0.5 mole of molecules		1 mole of molecules	and so on …
Or change moles to masses, using A_r and M_r	or	4 g 2 g		32 g 16 g		36 g 18 g	and so on …

Calculations from equations

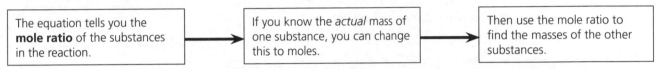

| The equation tells you the **mole ratio** of the substances in the reaction. | $\rightarrow$ | If you know the *actual* mass of one substance, you can change this to moles. | $\rightarrow$ | Then use the mole ratio to find the masses of the other substances. |

Example 1   In the complete combustion of methane (CH_4), what mass of oxygen combines with 64 g of methane, and how much carbon dioxide is produced?

Equation	$CH_4(g)$ $+$ $2O_2(g)$ $\rightarrow$ $CO_2(g)$ $+$ $2H_2O(l)$
Mole ratio	1 2 1 2
Moles of known substance	M_r of $CH_4 = 16$ so 64 g of $CH_4 = (64 \div 16)$ moles $= 4$ moles
Using the mole ratio	4 moles of CH_4 so 8 moles of O_2 and 4 moles of CO_2
Change moles to masses	M_r of $O_2 = 32$; $(8 \times 32) = $**256 g** M_r of $CO_2 = 44$; $(4 \times 44) = $**176 g**
So …	**256 g of oxygen** combines with 64 g of methane and produces **176 g of carbon dioxide**

A_r values
H = 1
C = 12
O = 16

Example 2   Aluminium burns in oxygen. What mass of oxygen combines with 100 g of aluminium, and how much aluminium oxide is produced?

Equation	$4Al(s)$ $+$ $3O_2(g)$ $\rightarrow$ $2Al_2O_3(s)$
Mole ratio	4 3 2 but aluminium is the known substance, so make it 1 in the ratios: 1 $3 \div 4 = 0.75$ $2 \div 4 = 0.5$
Moles of known substance	$100 \div 27 = 3.704$ moles of Al
Using the mole ratio	3.704 of Al so $(3.704 \times 0.75) = 2.778$ of O_2 and $(3.704 \times 0.5) = 1.852$ of Al_2O_3
Change moles to masses	O_2: $(2.778 \times 32) = $**88.9 g** Al_2O_3: $(1.852 \times 102) = $**188.9 g**
So …	**88.9 g of oxygen** combines with 100 g of aluminium and produces **188.9 g of aluminium oxide**

A_r values
O = 16
Al = 27

✔

Quick check for 6.3 (*Answers on page 165*)

1   How much oxygen is required for the complete combustion of 128 g of methane?
2   What mass of hydrogen will combine with 56 g of nitrogen in this reaction?
    $N_2 (g) + 3H_2 (g) \longrightarrow 2NH_3 (g)$
3   What mass of ammonia could be produced from 56 g of nitrogen?
4   Magnesium carbonate breaks down on heating, like this:
    $MgCO_3 (s) \longrightarrow MgO (s) + CO_2 (g)$
    When 10 g of magnesium carbonate is heated, what mass of each product forms?

A_r values
H = 1
C = 12
N = 14
O = 16
Mg = 24

6.4 Calculations about solutions

Concentration: the amount of a solute (in grams or moles) in 1 dm³ of solution.
Units used: g/dm³ or mol/dm³
Remember: 1 dm³ = 1 litre = 1000 cm³
    A solution containing 1 mole of a solute in 1 dm³ of solution is
    often called a **molar solution** or a **1M solution**.
    A 2M solution contains 2 moles in 1 dm³ of solution, and so on.

Use the calculation triangle!

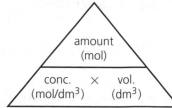

To find	Examples

concentration in mol/dm³	**concentration in mol/dm³** $= \dfrac{\textbf{amount of solute (mol)}}{\textbf{volume of solution (dm}^3\textbf{)}}$
	
	1 What is the concentration of a solution containing 2 moles of a compound in 0.5 dm³?
	    $2 \div 0.5 =$ **4 mol/dm³**
	2 What is the concentration of a solution containing 0.1 moles of a compound in 40 cm³?
	    $0.1 \div 0.04 =$ **2.5 mol/dm³**

concentration in g/dm³	**concentration in g/dm³ = concentration in mol/dm³ × M_r**
	1 What is the concentration in g/dm³ of a 4 mol/dm³ solution of sodium hydroxide? ($M_r = 40$)
	    4 mol/dm³ × 40 = **160 g/dm³**
	2 What is the concentration in g/dm³ of a 2.5 mol/dm³ solution of sulfuric acid? ($M_r = 98$)
	    2.5 mol/dm³ × 98 = **245 g/dm³**

volume of solution	**volume of solution (dm³)** $= \dfrac{\textbf{amount of solute (mol)}}{\textbf{concentration (mol/dm}^3\textbf{)}}$
	
	1 What volume of a 2 mol/dm³ solution contains 0.6 moles?
	    $0.6 \div 2 =$ **0.3 dm³** or **300 cm³**
	2 What volume of a 0.5 mol/dm³ solution contains 2 moles?
	    $2 \div 0.5 =$ **4 dm³** or **4000 cm³**

moles of solute	**amount of solute (mol) = concentration (mol/dm³) × volume of solution (dm³)**
	1 How many moles of solute are there in 2 dm³ of a 2 mol/dm³ solution?
	    2 × 2 = **4 moles**
	2 How many moles of solute are there in 50 cm³ of a 0.25 mol/dm³ solution?
	    0.25 × 0.05 = **0.0125 moles**

Calculating the volume of a solution in a reaction
You use the same logic as in section 6.3.

Example 1 What volume of 0.5 mol/dm³ hydrochloric acid reacts with 0.12 g of magnesium?

Equation	Mg (s) + 2HCl (aq) ⟶ MgCl₂ (aq) + H₂ (g)
Mole ratio	1 2
Moles of known substance	Mg: 0.12 ÷ 24 = 0.005 moles
Using the mole ratio	0.005 moles of Mg so (0.005 × 2) = 0.01 moles of HCl
Volume	volume = mol ÷ concentration (see the calculation triangle) so volume of HCl solution = (0.01 ÷ 0.5) = **0.02 dm³** or **20 cm³**
So ...	**20 cm³ of 0.5 mol/dm³ hydrochloric acid** reacts with 0.12 g of magnesium

A_r value
Mg = 24

Example 2 What volume of 2.0 mol/dm³ sodium hydroxide neutralises 25 cm³ of 0.5 mol/dm³ sulfuric acid?

Equation	2NaOH (aq) + H₂SO₄ (aq) ⟶ Na₂SO₄ (aq) + H₂O (l)
Mole ratio	1 2
Moles of known substance	H₂SO₄: 0.5 × 0.025 = 0.0125 moles
Using the mole ratio	0.0125 moles of H₂SO₄ so (2 × 0.0125) = 0.0025 moles of NaOH
Volume	volume = mol ÷ concentration (see the calculation triangle) so volume of NaOH solution = (0.0025 ÷ 2) = **0.00125 dm³** or **12.5 cm³**
So ...	**12.5 cm³ of 2 mol/dm³ sodium hydroxide** neutralises 25 cm³ of 0.5 mol/dm³ of sulfuric acid.

Quick check for 6.4   *(Answers on page 165)*
1 What is the concentration of a solution containing:
   a 0.25 moles in 2 dm³ of solution?   **b** 0.6 moles in 100 cm³ of solution?
2 What volume of a 0.5 mol/dm³ solution contains:
   a 2 moles?   **b** 0.01 moles?
3 Find the number of moles of solute in:
   a 3 dm³ of a 0.05 mol/dm³ solution   **b** 40 cm³ of a 2.5 mol/dm³ solution
4 How many grams of potassium hydroxide (M_r = 56) are there in:
   a 2 dm³ of a molar solution?   **b** 20 cm³ of a 0.5 mol/dm³ solution?
5 What volume of 2 mol/dm³ nitric acid will react with 10 g of calcium carbonate (M_r = 100)? The equation for the reaction is:
   CaCO₃ (s) + 2HNO₃ (aq) ⟶ Ca (NO₃)₂ (aq) + H₂O (l) + CO₂ (g)
6 What volume of 0.5 mol/dm³ sulfuric acid will neutralise 15 cm³ of sodium hydroxide solution, of concentration 1 mol/dm³?

Extended

6.5 Calculating volumes of gases

Remember

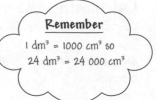

1 dm³ = 1000 cm³ so
24 dm³ = 24 000 cm³

Key idea: 1 mole of a gas has a volume of **24 dm³** at room temperature and pressure (rtp).
So 24 dm³ is called **the molar gas volume**.

To find	Examples

volume	**volume of gas at rtp (dm³) = no. of moles × 24 dm³**
	 1 What volume does 3 moles of a gas occupy, at rtp?
	        3 × 24 = **72 dm³** or **72 000 cm³**
	 2 What volume does 14 g of nitrogen (N_2) occupy, at rtp? (A_r: N =14)
	        Number of moles of nitrogen gas = 14 ÷ 28 = 0.5
	        Volume = 0.5 × 24 = **12 dm³** or **12 000 cm³**

moles	**no. of moles = volume of gas (dm³)**
	                                24 dm³
	 1 How many moles are there in 48 dm³ of a gas, at rtp?
	        48 ÷ 24 = **2 moles**
	 2 How many moles are there in 120 cm³ of a gas, at rtp?
	        120 ÷ 24000 = **0.005 moles**

Use the calculation triangle!

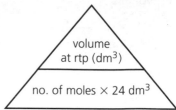

volume
at rtp (dm³)

no. of moles × 24 dm³

Calculating the volume of a gas in a reaction

Example 1   What volume of hydrogen forms, at rtp, when 0.12 g of magnesium reacts with hydrochloric acid?

Equation	Mg (s) + 2HCl (aq) $\longrightarrow$ $MgCl_2$ (aq) + H_2 (g)	
Mole ratio	1	1
Moles of known substance	A_r of Mg = 24 so (0.12 ÷ 24) = 0.005 moles of Mg	
Using the mole ratio	0.005 moles of Mg so 0.005 moles of H_2	
Volume at rtp	volume = no. of mol × 24 dm³ so volume of H_2 = (0.005 × 24) = **0.12 dm³** or **120 cm³**	
So ...	**120 cm³ of hydrogen** is released.	

Example 2   What volumes of nitrogen and hydrogen give 50 cm³ of ammonia gas, at rtp?
1 mole of every gas has the same volume at rtp (24 dm³).
So in reactions that involve only gases, we can say: **volume ratio = mole ratio**.

Equation	N_2 (g) + $3H_2$ (g) $\longrightarrow$ $2NH_3$ (g)		
Volume ratio = mole ratio	1	3	2
Using the volume ratio	25 cm³ of N_2	75 cm³ of H_2	50 cm³ of NH_3
So ...	**25 cm³ of nitrogen** and **75 cm³ of hydrogen** give 50 cm³ of ammonia.		

✓

Quick check for 6.5   *(Answers on page 165)*

1   Calculate the volume at rtp of:
      a 6 g of oxygen, O_2 **b** 6.4 g of sulfur dioxide, SO_2 (A_r: O =16, S = 32)
2   What volume of carbon dioxide at rtp is released when 5.3 g of sodium
      carbonate (M_r = 106) reacts with hydrochloric acid? The equation is:
      Na_2CO_3 (s) + 2HCl (aq) $\longrightarrow$ 2NaCl (aq) + H_2O (l) + CO_2 (g)
3   Hydrogen and chlorine react like this: H_2 (g) + Cl_2 (g) $\longrightarrow$ 2HCl (g)
      What volumes of hydrogen and chlorine give 250 dm³ of hydrogen chloride?

Extended

6.6 Calculating % yield and % purity

% yield
- The **yield** is the amount of product obtained from a chemical reaction.
- We can calculate the amount of product from the equation (the **calculated mass**).
- But the **actual mass** we get is less – for example because some reactant remains unreacted, or some product is lost in the separation process.

The % yield $= \dfrac{\text{actual mass}}{\text{calculated mass}} \times 100\%$

Example In an experiment, 100 g of aluminium is burnt in oxygen, giving aluminium oxide. 150 g of aluminium oxide is obtained. What is the % yield for the experiment?

Equation	$4Al\ (s)\ +\ 3O_2\ (g)\ \longrightarrow\ 2Al_2O_3\ (s)$
Mole ratio	4 3 2
Calculated mass	The calculated mass of aluminium oxide, from 100 g of Al, is 188.9 g. (See Example 2 on page 41 for the working for this.)
Actual mass	150 g of aluminium oxide
% yield	(150 ÷ 188.9) × 100 = 79.4 %
So ...	**the % yield for the experiment was 79.4 %.**

% purity
- When we carry out a reaction, we obtain a certain mass of product.
- But it is impure. There may still be some reactant mixed with it, for example.

The % purity of the product from the reaction $= \dfrac{\text{mass of pure product}}{\text{mass of impure product}} \times 100\ \%$

Example Aluminium was burned in oxygen, to give aluminium oxide. The equation is:
$$4Al\ (s)\ +\ 3O_2\ (g)\ \longrightarrow\ 2Al_2O_3\ (s)$$
150 g of product was obtained. But it was found to contain 5 g of impurities. What is the % purity of the aluminium oxide obtained?

Mass of impure product obtained	150 g
Mass of pure product	150 g – 5 g = 145 g
% purity	(145 ÷ 150) × 100 = **96.7%**
So ...	**the aluminium oxide obtained was 96.7% pure.**

✓
Quick check for 6.6 *(Answers on page 165)*
1   What would be the ideal yield, in an experiment?
2   According to the equation, a reaction should give you 60 g of product. But you obtain only 45 g. What is the % yield?
3   You obtain 7.5 g of compound X, in an experiment. After purifying, its mass is 6 g. What was the % purity of X, in your experiment?
4   Some sea water is evaporated. The salt obtained is 91% sodium chloride. How much sodium chloride would you obtain from 20 tonnes of this salt?
5   Electrolysis of 102 kg of aluminium oxide gave 45 kg of aluminium. What was the % yield? (A_r: Al = 27, O = 16)

Questions on section 6

Answers for these questions are on page 165.

Extended curriculum

1 Marble reacts with dilute hydrochloric acid according to the equation:

$$CaCO_3 \ (s) + 2HCl \ (aq) \longrightarrow CaCl_2 \ (aq) + CO_2 \ (aq) + H_2O \ (l)$$

One piece of marble, 0.3 g, was added to 5 cm³ of hydrochloric acid, 1.00 mol/dm³.

  a   Which reagent is in excess? Give a reason for your choice.
      mass of one mole of CaCO₃ = 100 g

  b   Use your answer to **a** to calculate the maximum volume of carbon dioxide
      produced, measured at rtp.
      The volume of one mole of any gas is 24 dm³ at room temperature and
      pressure (rtp).   *CIE 0620 November '07 Paper 3 Q7b ii and iii*

2 Crystals of sodium sulfate-10-water, $Na_2SO_4.10H_2O$, are prepared by titration.

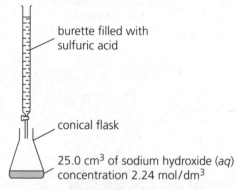

burette filled with
sulfuric acid

conical flask

25.0 cm³ of sodium hydroxide (*aq*)
concentration 2.24 mol/dm³

  a   25.0 cm³ of aqueous sodium hydroxide is pipetted into a conical flask.
      A few drops of an indicator are added. Using a burette, dilute sulfuric acid is
      slowly added until the indicator just changes colour. The volume of acid needed to
      neutralise the alkali is noted.
      Suggest how you would continue the experiment to obtain pure, dry crystals of
      sodium sulfate-10-water.

  b   Using 25.0 cm³ of aqueous sodium hydroxide, 2.24 mol/dm³, 3.86 g of crystals
      were obtained.

$$2NaOH + H_2SO_4 \longrightarrow Na_2SO_4 + 2H_2O$$
$$Na_2SO_4 + 10 \, H_2O \longrightarrow Na_2SO_4.10H_2O$$

      Calculate:

      i   The number of moles of NaOH used.
      ii   The maximum number of moles of $Na_2SO_4.10H_2O$ that could be formed.
      iii The maximum yield of sodium sulfate-10-water.
           The mass of one mole of Na₂SO₄.10H₂O = 322 g.
      iv The percentage yield.   *CIE 0620 June '08 Paper 3 Q7*

3 The results of an investigation into the action of heat on copper(II) sulfate-5-water, a
blue crystalline solid, are given below.
The formula is $CuSO_4.5H_2O$ and the mass of one mole is 250 g.
A 5.0 g sample of the blue crystals is heated to form 3.2 g of a white powder.
With further heating this decomposes into a black powder and sulfur trioxide.

  a   Name the white powder.
  b   What is observed when water is added to the white powder?
  c   Name the black powder.
  d   Calculate the mass of the black powder. Show your working.
                  CIE 0620 November '02 Paper 3 Q1

A_r values:
H = 1
O = 16
Cu = 64

Extended

Extended

4 Chemists use the concept of the mole to calculate the amounts of chemicals involved in a reaction.

    a Define *mole*.

    b 3.0 g of magnesium was added to 12.0 g of ethanoic acid.

$$Mg + 2CH_3COOH \longrightarrow (CH_3COO)_2Mg + H_2$$

    The mass of one mole of Mg is 24 g.

    The mass of one mole of CH_3COOH is 60 g.

      i Which one, magnesium or ethanoic acid, is in excess? You must show your reasoning.

      ii How many moles of hydrogen were formed?

      iii Calculate the volume of hydrogen formed, measured at rtp.

    c In an experiment, 25.0 cm³ of aqueous sodium hydroxide, 0.4 mol/dm³, was neutralised by 20.0 cm³ of aqueous oxalic acid, $H_2C_2O_4$.

$$2NaOH + H_2C_2O_4 \longrightarrow Na_2C_2O_4 + 2H_2O$$

    Calculate the concentration of the oxalic acid in mol/dm³, by these steps.

      i Calculate the number of moles of NaOH in 25.0 cm³ of 0.4 mol/dm³ solution.

      ii Use your answer to **i** and the mole ratio in the equation to find out the number of moles of $H_2C_2O_4$ in 20 cm³ of solution.

      iii Calculate the concentration, in mol/dm³, of the aqueous oxalic acid.

CIE 0620 June '04 Paper 3 Q2

5 The following method is used to make crystals of hydrated nickel sulfate.

An excess of nickel carbonate, 12.0 g, was added to 40 cm³ of sulfuric acid, 2.0 mol/dm³. The unreacted nickel carbonate was filtered off and the filtrate evaporated to obtain the crystals.

$$NiCO_3 + H_2SO_4 \longrightarrow NiSO_4 + CO_2 + H_2O$$
$$NiSO_4 + 7H_2O \longrightarrow NiSO_4.7H_2O$$

Mass of one mole of $NiSO_4.7H_2O$ = 281 g

Mass of one mole of $NiCO_3$ = 119 g

    a Calculate the mass of unreacted nickel carbonate, by these steps.

    Number of moles of H_2SO_4 in 40 cm³ of 2.0 mol/dm³ acid = 0.08

      i Number of moles of $NiCO_3$ reacted = …………

      ii Mass of nickel carbonate reacted = …………

      iii Mass of unreacted nickel carbonate = …………

    b The experiment produced 10.4 g of hydrated nickel sulfate. Calculate the percentage yield, by these steps.

      i The maximum number of moles of $NiSO_4.7H_2O$ that could be formed = ……

      ii The maximum mass of $NiSO_4.7H_2O$ that could be formed = ……

      iii The percentage yield = …… % *CIE 0620 November '05 Paper 3 Q6*

6 Two gases react as shown.

$$X_2 + Y_2 \longrightarrow 2XY$$

For this reaction, what is the missing number below?

$$\frac{\text{volume of product}}{\text{total volume of reactants}} = …… \text{ at rtp}$$

7 Electricity and chemical change

The big picture

- Some substances allow electricity to pass through them.
- When you pass electricity through a liquid that contains ions, decomposition (breaking down) takes place.
- The process of using electricity to break down a substance is called **electrolysis**.
- Electrolysis has many important uses in industry.

7.1 Conductors and insulators

- If a substance contains charged particles (electrons or ions) that are free to move, then electricity can pass through it. The moving particles carry the electric current.
- Substances that let electricity pass through them are called **conductors**. Substances that don't are called **non-conductors** or **insulators**.
- You can test a substance to see if it conducts by putting it into a circuit, as on the right.

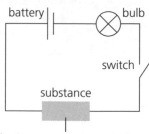

if substance conducts, bulb lights when switch is closed

Examples

Conductors	Why?
Copper wire Molten (melted) copper Carbon, in the form of graphite	They contain electrons that are free to move. *The substances themselves remain unchanged.*
Molten sodium chloride A solution of sodium chloride in water	They contain ions that are free to move. *But when the ions move, the substances break down or decompose.*

Non-conductors	Why?
Solid sodium chloride	The ions in the solid are held in a lattice by strong forces, so are not free to move.
Ceramics	Any ions in these solids are held in a lattice by strong forces, and not free to move.
Plastics	No charged particles free to move.

Making use of some conductors and insulators

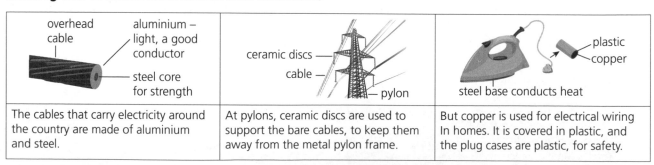

The cables that carry electricity around the country are made of aluminium and steel.	At pylons, ceramic discs are used to support the bare cables, to keep them away from the metal pylon frame.	But copper is used for electrical wiring In homes. It is covered in plastic, and the plug cases are plastic, for safety.

✓

Quick check for 7.1 *(Answers on page 165)*

1 Will it conduct electricity? Give a reason.
   a an iron rod   **b** a solution of sodium chloride in water (an aqueous solution)
   c an aqueous solution of sugar   **d** melted sugar   **e** mercury
2 **a** Solid potassium chloride is a non-conductor. Why?
   b Suggest two things you could do to potassium chloride, to make it conduct.

7.2 Electrolysis

- Liquids that contain ions will conduct electricity. But at the same time, decomposition takes place.
- So you can use electricity on purpose, to decompose (break down) a substance. The process is called electrolysis, which means *splitting by electricity*.

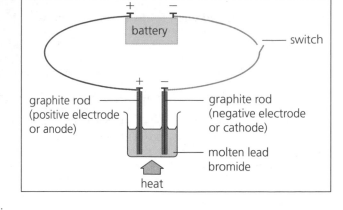

An example of electrolysis

Let's take the electrolysis of molten lead bromide as example. The electrolysis is carrried out using the apparatus on the right. The graphite rods are called **electrodes**. They carry the current into and out of the liquid.

The drawing below shows what happens when the switch is closed:

5 Electrons flow from the anode back to the positive terminal of the battery.

1 Electrons flow this way, from the negative terminal of the battery to the cathode.

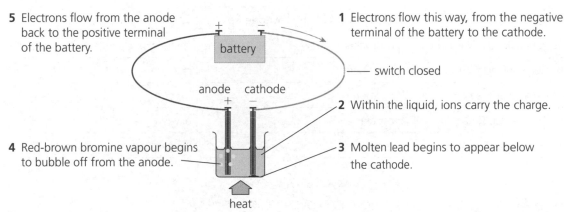

2 Within the liquid, ions carry the charge.

4 Red-brown bromine vapour begins to bubble off from the anode.

3 Molten lead begins to appear below the cathode.

Note

- Electrodes made of carbon (graphite) or platinum are **inert**: they carry current, but are not themselves changed during electrolysis.
- The **anode** is the positive electrode (+), joined to the positive terminal of the battery. The **cathode** is the negative electrode (–).
- The liquid is called the **electrolyte**.
- Electrons carry the current in the wires; ions carry it in the liquid.

> **Remember**
> PANCAKE –
> Positive Anode,
> Negative Cathode

The products of electrolysis

Electrolysis always gives a metal or hydrogen at the cathode (–), and a non-metal other than hydrogen at the anode (+).

Look at the following examples.

1   Starting with a molten compound

The electrolysis of a molten ionic compound always gives a metal at the cathode, and a non-metal at the anode. For example:

The molten compound	What forms at the cathode (–)	What forms at the anode (+)	The decomposition
lead bromide	lead	bromine	lead bromide $\rightarrow$ lead + bromine $PbBr_2\,(l) \quad \rightarrow \quad Pb\,(l) + Br_2\,(g)$
lead iodide	lead	iodine	lead iodide $\rightarrow$ lead + iodine $PbI_2\,(l) \quad \rightarrow \quad Pb\,(l) + I_2\,(g)$
sodium chloride	sodium	chlorine	sodium chloride $\rightarrow$ sodium + chlorine $2\,NaCl\,(l) \quad \rightarrow \quad 2\,Na\,(l) + Cl_2\,(g)$

2   Starting with a solution

The electrolysis of aqueous solutions is more complex, because the water also plays a part. These are the rules:

<table>
<tr><td colspan="2">**A metal OR hydrogen forms at the cathode (–).**</td></tr>
<tr><td>**1**</td><td>If the metal is more reactive than hydrogen, hydrogen forms. The metal ions stay in solution. (Check the reactivity series on the right.)</td></tr>
<tr><td>**2**</td><td>If the metal is less reactive than hydrogen, the metal forms.</td></tr>
<tr><td colspan="2">**A non-metal other than hydrogen forms at the anode (+).**</td></tr>
<tr><td>**1**</td><td>A concentrated solution of a halide (a chloride, bromide or iodide) gives chlorine, bromine or iodine at the anode.</td></tr>
<tr><td>**2**</td><td>But if there is no halide present, or the halide solution is dilute, oxygen forms instead.</td></tr>
</table>

The reactivity series	
potassium	
sodium	
calcium	
magnesium	
aluminium	
zinc	reactivity increases
iron	
lead	
hydrogen	
copper	
silver	

Look at these examples:

Solution	What forms at the cathode (–)	What forms at the anode (+)
concentrated sodium chloride, NaCl (*aq*)	hydrogen	chlorine
concentrated hydrochloric acid, HCl (*aq*)	hydrogen	chlorine
dilute sodium chloride, NaCl (*aq*)	hydrogen	oxygen
dilute copper(II) chloride, $CuCl_2$ (*aq*)	copper	oxygen
copper(II) sulfate, $CuSO_4$ (*aq*)	copper	oxygen (not sulfur dioxide!)

✓

Quick check for 7.2   *(Answers on page 165)*

1   What is: **a** an electrode?   **b** a electrolyte?
2   **a** How can you tell which electrode is positive?
       b Give another name for this electrode.
3   Give a general rule for the electrolysis of molten ionic compounds.
4   What will you get when you pass an electric current through:
       a molten potassium chloride?   **b** a dilute solution of potassium chloride?

7.3 A closer look at the electrode reactions

During electrolysis, the ions move to the electrode of opposite charge. (That is how the current is carried through the liquid.) Now look what happens to them at the electrodes:

Remember

RAC: Reduction At Cathode.

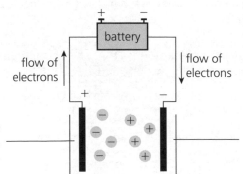

At the anode (+) the negative ions give up electrons and become atoms. This is **oxidation**.

At the cathode (–) the positive ions accept electrons and become atoms. This is **reduction**.

So electrolysis is a redox reaction.
Oxidation takes place at the anode, and reduction at the cathode.

Extended

Extended

Half equations for electrode reactions

We use **half equations** to show the reactions at the electrodes.
One half equation shows electron loss, the other shows electron gain.

> **Remember**
> OILRIG: Oxidation Is Loss,
> Reduction Is Gain
> (of electrons)

Example 1: molten lead bromide		
Ions present lead ions, Pb^{2+} bromide ions, Br^-	**Reduction at the cathode (–)** each lead ion accepts two electrons and becomes a lead atom. The half equation is: Pb^{2+} (*l*) + $2e^-$ → Pb (*l*)	**Oxidation at the anode (+)** each bromide ion gives up an electron, and becomes an atom. Atoms join to form molecules. The half equation is: $2Br^-$ (*l*) → Br_2 (*g*) + $2e^-$
So the overall reaction is the decomposition of lead bromide: $PbBr_2$ (*l*) → Pb (*l*) + Br_2 (*g*)		

Example 2: a concentrated solution of sodium chloride		
This time, water is present, and that makes a difference – because in water, a tiny % of the molecules is always broken down into ions: H_2O (*l*) $\xrightarrow{\text{a tiny \% of molecules}}$ H^+ (*aq*) + OH^- (*aq*)		
Ions present Na^+ Cl^- from sodium chloride H^+ OH^- from water	**Reduction at the cathode (–)** the H^+ ions accept electrons: $2H^+$ (*aq*) + $2e^-$ → H_2 (*g*)	**Oxidation at the anode (+)** the Cl^- ions give up electrons: $2Cl^-$ (*aq*) → Cl_2 (*g*) + $2e^-$
So the overall reaction is: $2NaCl$ (*aq*) + $2H_2O$ (*l*) → $2NaOH$ (*aq*) + Cl_2 (*g*) + H_2 (*g*)		

Example 3: concentrated hydrochoric acid		
Ions present H^+ Cl^- from hydrochloric acid H^+ OH^- from water	**At the cathode (–)** as in Example 2	**At the anode (+)** as in Example 2
So the overall reaction is the decomposition of hydrochloric acid: $2HCl$ (*aq*) → H_2 (*g*) + Cl_2 (*g*)		

✓

Quick check for 7.3 *(Answers on page 165)*

1 Write the half equations for the reactions at the electrodes, during the electrolysis of molten sodium chloride.

2 A solution of potassium chloride contains some hydroxide ions. Explain why.

> **Note**
> A half equation is also called the **ionic equation** for the reaction at an electrode.

7.4 Different electrodes, different products

Extended

Copper(II) sulfate solution contains blue Cu^{2+} ions, and SO_4^{2-} ions, and H^+ and OH^- ions from water. Compare these two electrolyses of a copper(II) sulfate solution.

Electrodes	At the cathode (–)	At the anode (+)	Comment
graphite	copper forms: $2Cu^{2+}$ (*aq*) + $4e^-$ → 2Cu (*s*)	oxygen bubbles off: $4OH^-$ (*aq*) → $2H_2O$ (*l*) + O_2 (*g*) + $4e^-$	The electrodes are inert – they are not changed. The products fit the rules on page 50.
copper	again copper forms: Cu^{2+} (*aq*) + $2e^-$ → Cu (*s*)	the anode **dissolves**: Cu (*s*) → Cu^{2+} (*aq*) + $2e^-$	This time the anode is changed by the electrolysis.

• So changing the electrodes can change the products of electrolysis.
• The idea of dissolving anodes is used in **electroplating**, and in **refining copper**.

✓
Quick check for 7.4   *(Answers on page 165)*
1   When a solution of copper(II) sulfate is electrolysed using graphite electrodes,
    its blue colour fades. Why?
2   But when copper electrodes are used instead, explain why:
    a the blue colour does not change
    b the anode loses mass, while the cathode gains the same amount of mass.

7.5 Four uses of electrolysis in industry

1   Electroplating

- In electroplating, one metal is coated with another to protect it against corrosion,
  or make it look more attractive. For example:
  - ➤   steel is coated with tin to make rust-proof tin cans for food
  - ➤   steel taps or car bumpers are coated with chromium, to make them look shiny.
- To electroplate a metal object with a coat of metal X:
  use metal X as the anode (+)
  use the object as the cathode (–)
  use a solution of a soluble compound of X as the electrolyte.

> **Remember**
> **MOC:** **M**ake the
> **O**bject the **C**athode.
> for electroplating

Here, a metal jug is being electroplated with a coat of silver, using an anode of silver:

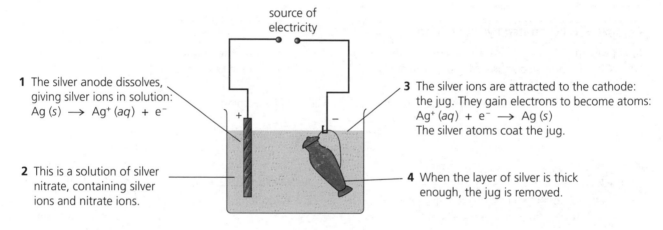

source of electricity

1 The silver anode dissolves,
giving silver ions in solution:
$Ag\ (s) \longrightarrow Ag^+\ (aq) + e^-$

2 This is a solution of silver
nitrate, containing silver
ions and nitrate ions.

3 The silver ions are attracted to the cathode:
the jug. They gain electrons to become atoms:
$Ag^+\ (aq) + e^- \longrightarrow Ag\ (s)$
The silver atoms coat the jug.

4 When the layer of silver is thick
enough, the jug is removed.

2   Refining copper

Electrolysis is used to purify or **refine** copper, for electrical wiring. (The purer the copper is,
the better it conducts.) The refining process works in the same way as electroplating:

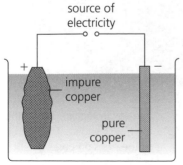

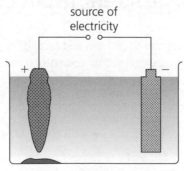

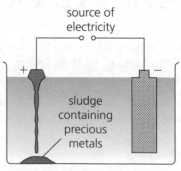

Anode: impure copper. Cathode: pure copper. Electrolyte: dilute copper(II) sulfate solution.	The copper in the anode dissolves. The impurities fall to the cell floor as a sludge. A layer of pure copper builds up on the cathode.	When the anode is almost gone, the anode and cathode are replaced. Precious metals are removed from the sludge and sold.

Extended

3 Manufacturing useful chemicals from brine

- Brine is a concentrated solution of common salt (sodium chloride). The ions present are: Na^+, Cl^-, H^+, OH^-
- The electrolysis follows the rules on page 50 …
    - ➤ **chlorine** forms at the anode, and is collected: $2Cl^- \longrightarrow Cl_2 + 2e^-$
    - ➤ **hydrogen** forms at the cathode, and is collected: $2H^+ + 2e^- \longrightarrow H_2$
    - ➤ the sodium and hydroxide ions remain in solution, giving a solution of **sodium hydroxide**. Evaporating the water gives solid sodium hydroxide.
- The products have many uses. Look at the panel on the right.

4 Extracting metals from their ores

- Metal ores are compounds that occur naturally in the Earth. They can be dug up and the metal extracted from them.
- The more reactive the metal, the harder it is to break down its compounds.
- Electrolysis is a powerful way to break down metal compounds, to get the metal. So it is used to extract the more reactive metals such as sodium and aluminium from their compounds.

> **Uses for chemicals from brine**
>
> **chlorine** – for making PVC, medical drugs, bleaches, paints, and dyes
>
> **sodium hydroxide** – for making soaps, detergents, medical drugs, dyes
>
> **hydrogen** – for making ammonia, margarines, and as a fuel

Make the link to … methods of extraction, on page 117.

Example: extracting aluminium

The starting material is the aluminium ore **bauxite**:

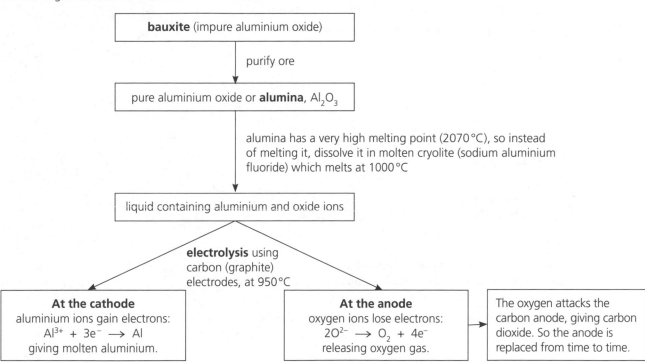

> **bauxite** (impure aluminium oxide)
>
> purify ore
>
> pure aluminium oxide or **alumina**, Al_2O_3
>
> alumina has a very high melting point (2070 °C), so instead of melting it, dissolve it in molten cryolite (sodium aluminium fluoride) which melts at 1000 °C
>
> liquid containing aluminium and oxide ions
>
> **electrolysis** using carbon (graphite) electrodes, at 950 °C
>
> **At the cathode**
> aluminium ions gain electrons:
> $Al^{3+} + 3e^- \longrightarrow Al$
> giving molten aluminium.
>
> **At the anode**
> oxygen ions lose electrons:
> $2O^{2-} \longrightarrow O_2 + 4e^-$
> releasing oxygen gas.
>
> The oxygen attacks the carbon anode, giving carbon dioxide. So the anode is replaced from time to time.

The overall reaction is the decomposition of aluminium oxide:

$$2Al_2O_3 \, (l) \longrightarrow 4Al \, (l) + 3O_2 \, (g)$$

✓
Quick check for 7.5 *(Answers on page 165)*

1 You want to electroplate a steel spoon with nickel. How will you do it?
2 In what ways are electroplating, and the refining of copper, similar?
3 Explain why brine is used instead of a dilute solution of sodium chloride, to manufacture chlorine.
4 Aluminium oxide melts at over 2000 °C. So it would be too difficult to melt it, for electrolysis. How do they overcome this problem?
5 Write the ionic equations for the electrode reactions, in extracting aluminium.

Questions on section 7

Answers for these questions start on page 165.

Core curriculum

1   The diagram shows the structure of lead bromide.

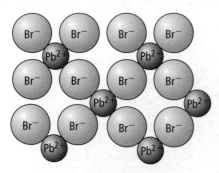

a   What is the simplest formula for lead bromide?

b   What type of structure and bonding is present in lead bromide?
Choose two words from the following:

atomic   covalent   giant   ionic   metallic   molecular

c   Lead bromide is electrolysed using the apparatus shown below.

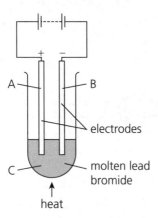

i   Which letter, A, B or C, represents the cathode?

ii   State the name of a metal which can be used for the electrodes.

iii   Why does lead bromide have to be molten for electrolysis to occur?

iv   State the name of the products formed at the anode and cathode during
this electrolysis.   *CIE 0620 November '06 Paper 2 Q6*

2   The diagram shows how iron is electroplated with chromium.

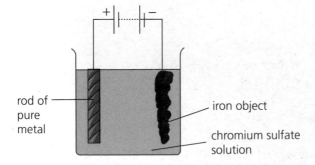

a   Is the iron object the anode or cathode, for the electroplating?

b   What is the purpose of the chromium sulfate solution?

c   Name the pure metal that must be used for the electroplating.

3 Copper can be purified by electrolysis.

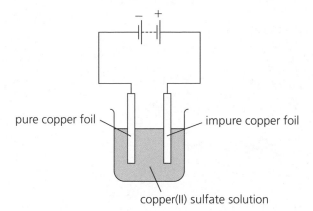

pure copper foil — impure copper foil

copper(II) sulfate solution

a Choose a word from the list below which describes the pure copper foil.

anion   anode   cathode   cation   electrolyte

b Describe what happens during this electrolysis to

i the pure copper foil

ii the impure copper foil.   *CIE 0620 June '08 Paper 2 Q7*

Extended curriculum

1 Ionic compounds such as lithium chloride and copper(II) chloride can be electrolysed as long as the ions are free to move.

a State the two ways in which the ions can be made free to move.

b Explain how the ions move, when an ionic compound is electrolysed with platinum electrodes.

c Write ionic equations for the reaction at: **i** the anode   **ii** the cathode in the electrolysis of a concentrated solution of copper(II) chloride.

d i The anode reaction is different if the solution is dilute. Explain why.

ii Write the equation for the anode reaction using a dilute solution.

e Explain why copper is obtained at the cathode, not lithium.

f i What will happen if the platinum anode is replaced by copper?

ii Write the equation that will take place at this anode.

iii What type of reaction is this?

2 This is about the electrolysis of hydrochloric acid.

a Explain the term electrolysis.

b Give the names and symbols of all the ions present in an aqueous solution of hydrochloric acid.

c i Say what will be formed at the negative electrode (cathode), and write an ionic equation for the reaction at the electrode.

ii Is this reaction an oxidation, or a reduction? Explain your answer.

d The product formed at the positive electrode (anode) depends on the concentration of the hydrochloric acid. What will the product be when the hydrochloric acid is:

i dilute?   **ii** concentrated?

3 The electrolysis of concentrated aqueous sodium chloride produces three commercially important chemicals: hydrogen, chlorine and sodium hydroxide.

a The ions present are Na^+ (*aq*), H^+ (*aq*), Cl^- (*aq*) and OH^- (*aq*).

i Write the ionic equation for the reaction at the negative electrode (cathode).

ii Write the ionic equation for the reaction at the positive electrode (anode).

iii Explain why the solution changes from sodium chloride to sodium hydroxide.

b i Why does the water supply industry use chlorine?

ii Name an important chemical that is made from hydrogen.

iii Which chemical is used to make soap?   *CIE 0620 November '08 Paper 3 Q5*

4 Aluminium is extracted by the electrolysis of aluminium oxide dissolved in cryolite.

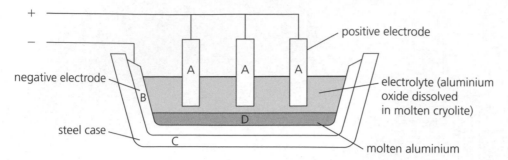

a Name the main ore of aluminium.

b Why does the molten electrolyte contain cryolite?

c What information in the diagram shows that aluminium is more dense than the electrolyte?

d What form of carbon is used for the electrodes in this electrolysis?

e Which letter in the diagram, A, B, C or D, represents the anode?

f Suggest why electrolysis is used to extract aluminium, rather than reduction using carbon.

g Oxygen gas is released at the anode.

   i Where does this oxygen come from?

   ii The oxygen reacts with the carbon anode to form carbon dioxide. What is the formula of carbon dioxide?

   iii Why does the anode decrease in size during electrolysis?

h Each electrolysis cell makes 212 kg of aluminium per day from 400 kg of aluminium oxide. Calculate how much aluminium can be made from 1 tonne (1000 kg) of aluminium oxide.

i Complete the following sentences about the electrolysis of aluminium oxide using words from the following list.

   atoms   gaseous   molten   solid   ions   molecules

   Aluminium oxide conducts electricity when it is (i)...
   because it contains (ii).. which are free to move.

   CIE 0620 June '07 Paper 2 Q7

Alternative to practical

1 Lead bromide was placed in a tube and connected to an electrical circuit as shown below.

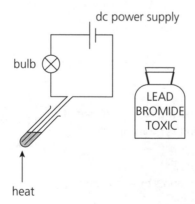

The lead bromide was heated until molten. A brown gas was given off.

   a State one other expected observation.

   b Suggest a suitable material for the electrodes.

   c Name the brown gas. At what electrode will the gas be given off?

   d Why is this experiment carried out in a fume cupboard?

   CIE 0620 June '04 Paper 6 Q3

8 Energy changes in reactions

The big picture

- There is always an overall energy change, during a reaction.
- Some reactions give out energy, overall. Some take it in.
- When we use fuels, energy is given out, overall, mainly in the form of heat. (That's why we use them!)
- We use some reactions to give energy in the form of electricity.

8.1 Exothermic and endothermic reactions

- During all chemical reactions, energy is given out or taken in, overall.
- Some chemical reactions are **exothermic**: they give out energy.
- The others are **endothermic**: they take in energy.

> **Remember**
> Exo means out (like exit).
> So exothermic means
> energy is given out.

	Exothermic reaction	Endothermic reaction
Definition	energy given out	energy taken in
Example	the reaction between hydrochloric acid and magnesium	the reaction between sodium hydrogen carbonate (baking soda) and citric acid
How the energy change is observed	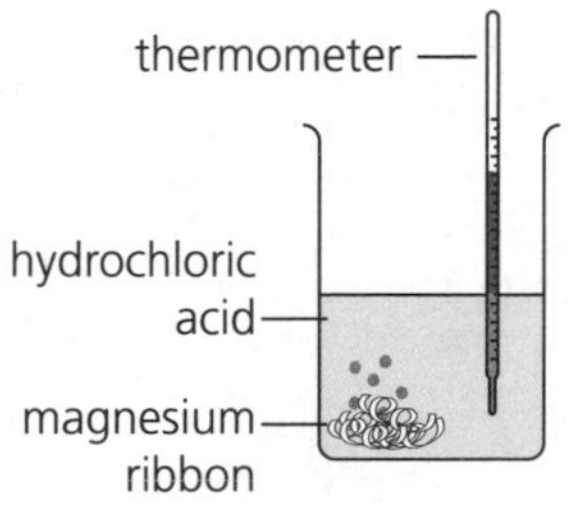 There is a rise in temperature.	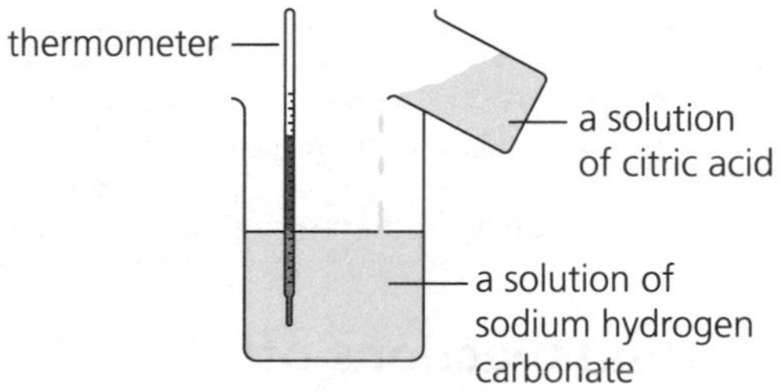 There is a fall in temperature.
Showing the energy change on an energy level diagram	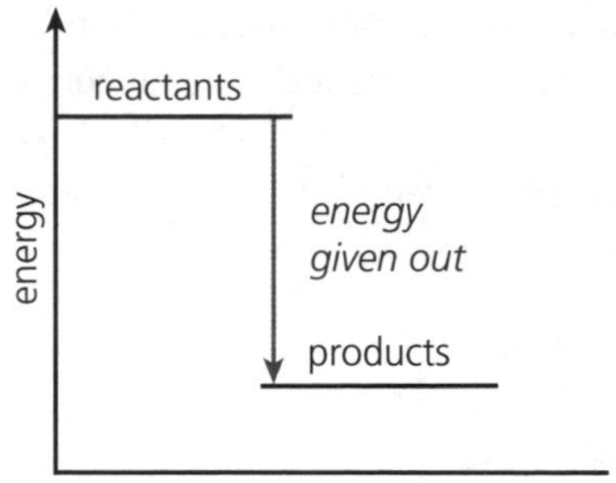 The products have **lower** energy than the reactants.	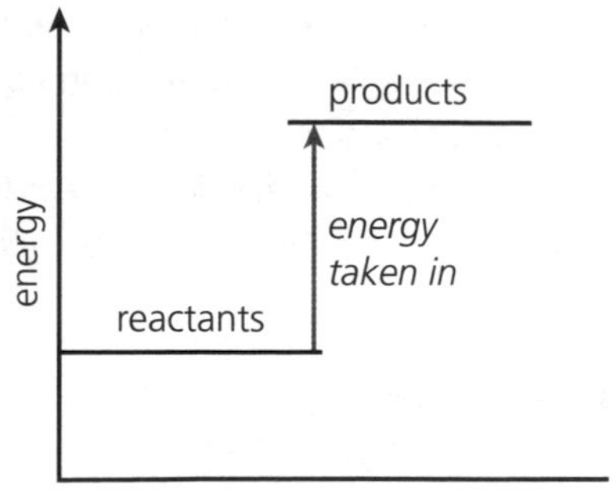 The products have **higher** energy than the reactants.
More examples of this type of reaction	• the neutralisation of an acid by an alkali • the combustion of fuels • the reaction between iron and sulfur	• the thermal decomposition of calcium carbonate • photosynthesis (uses energy from sunlight) • reactions that take place during cooking
Unit of measurement	The energy given out or taken in is measured in **kilojoules** (kJ).	

Why is there is an energy change?

In a chemical reaction, bonds are broken, and new bonds are formed.

Look at the reaction between hydrogen and chlorine, to give hydrogen chloride:

$H_2 + Cl_2 \longrightarrow 2HCl$

Reactants	⟶	Product

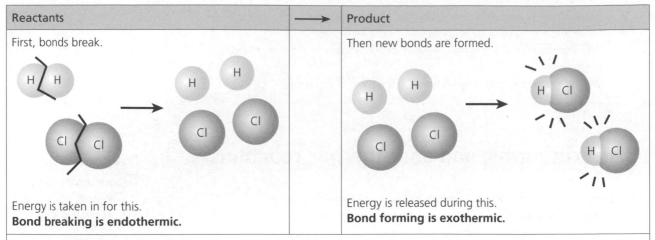

First, bonds break.

Energy is taken in for this.
Bond breaking is endothermic.

Then new bonds are formed.

Energy is released during this.
Bond forming is exothermic.

The amount of energy needed to break bonds is different from the amount released in forming new bonds.
That is why there is an overall energy change, in a reaction.

The overall energy change

Energy is taken in to make bonds, and given out when new bonds form.

So what is the overall energy change in the reaction?

If ...	then ...	so ...
more energy is taken in than given out	the reaction absorbs energy, overall	it is an endothermic reaction
less energy is taken in than given out	the reaction releases energy, overall	it is an exothermic reaction

Starting reactions off

*   Some reactions are **spontaneous**: they begin as soon as the reactants are mixed.
    The reaction between magnesium and hydrochloric acid is an example.
*   Some exothermic reactions need heat to start them off. For example magnesium must
    be lit with a match or Bunsen burner. This provides the energy to break enough bonds
    to start the reaction.
*   For an endothermic reaction such as the thermal decomposition of calcium carbonate,
    you need to keep heating until the reaction is over.

✓

Quick check for 8.1   *(Answers on page 166)*

1   In an exothermic reaction, energy is …? Complete the sentence.
2   Draw an energy level diagram for an endothermic reaction.
3   Magnesium burns in oxygen to form magnesium oxide. Which has lower energy:
    magnesium and oxygen (the reactants) or magnesium oxide (the product)?
4   Does making bonds absorb energy, or release energy?
5   Nitrogen and hydrogen react to give ammonia, in an exothermic reaction.
    Ammonia can break down again to nitrogen and hydrogen.
    a Is this reaction of ammonia exothermic, or endothermic?
    b Explain your answer in terms of bond breaking and bond making.

8.2 Calculating the energy change

Extended

Every bond between two atoms has a bond energy value. You can use these values to calculate the overall energy change in a reaction.

For the calculations
- Make sure you use the balanced equation for the reaction.
- Draw out the equation, showing all the bonds as lines, to help you.
- Remember that the calculation is always this way round …
  energy change = energy in to break bonds – energy out in forming bonds.
- Don't forget to include the sign (+ or –) in your answer: **a minus sign (–) shows an exothermic reaction, a plus sign (+) shows an endothermic one.**

Example 1: The reaction between hydrogen and chlorine

$$H-H + Cl-Cl \longrightarrow 2\,H-Cl$$

Energy in, to break each mole of bonds:

$1 \times H-H$	436 kJ
$1 \times Cl-Cl$	242 kJ
Total energy in	678 kJ

Energy out, from two moles of bonds forming:

$2 \times H-Cl$	$2 \times 431 = 862$ kJ

Energy in – energy out $= 678$ kJ $- 862$ kJ $= $ **–184 kJ**

So the reaction gives out 184 kJ of energy, overall. It is an exothermic reaction.

Example 2: The decomposition of ammonia

$$2\ \overset{\displaystyle H}{\underset{\displaystyle H}{N}}-H \longrightarrow N\equiv N + 3\,H-H$$

Energy in, to break the two moles of bonds:

$6 \times N-H$	$6 \times 391 = 2346$ kJ

Energy out, from four moles of bonds forming:

$1 \times N\equiv N$	946 kJ
$3 \times H-H$	$3 \times 436 = 1308$ kJ
Total energy out	2254 kJ

Energy in – energy out $= 2346$ kJ $- 2254$ kJ $= $ **+92 kJ**

So the reaction takes in 92 kJ of energy, overall. It is an endothermic reaction.

✓

Quick check for 8.2 (*Answers on page 166*)

1 One mole of hydrogen chloride is obtained by reacting hydrogen and chlorine.
   How much energy is given out in the reaction? (Check Example 1 above.)

2 When natural gas (methane) burns in air, the reaction is:
   $CH_4\,(g) + 2O_2\,(g) \longrightarrow CO_2\,(g) + 2H_2O\,(l)$
   a Do a drawing showing all the bonds. (Carbon dioxide has two double bonds!)
   b Calculate the energy needed to break all the bonds in the reactants.
   c Then calculate the energy given out when the new bonds form.
   d Is this reaction endothermic, or exothermic?
   e Give the energy change, with the correct sign and units.

Bond energies	(kJ/mole)
C–H	413
O=O	498
C=O	805
O–H	464

8.3 Using fuels to provide energy

- A **fuel** is a substance that we use as a source of energy.
- The energy is usually in the form of heat, which we use for cooking, heating, driving cars, and making steam to generate electricity.
- Many fuels must be burned, to release their energy – but not all. Look at the table below.
- The amount of energy given out when 1 mole of a fuel burns completely in oxygen is called the **heat of combustion**.

Comparing two kinds of fuels

	Fuels we burn	Nuclear fuels
The fuels	The fossil fuels: oil (petroleum), natural gas, coal.	Radioisotopes such as uranium-235; these are unstable atoms which break down (page 17)
The key idea	Fossil fuels burn in the oxygen in air.	Nuclear fuels are not burned; instead the atoms are broken down by bombarding them with neutrons.
Example	Natural gas (methane) burns in air, giving carbon dioxide and water: $CH_4 (g) + 2O_2 (g) \rightarrow CO_2 (g) + 2H_2O (l)$ + energy	Uranium-235 breaks down to give to smaller atoms such as lanthanum and bromine atoms: U-235 $\rightarrow$ La-145 + Br-88 + energy
The change	Chemical energy is released as heat.	Atomic energy is released as heat.
How we use this reaction	Natural gas is burned: - in homes, for cooking and heating - in gas-fired power stations; the heat is used to boil water to make steam, which spins turbines for generating electricity.	In nuclear power stations; the heat is used to boil water to make steam, which spins turbines for generating electricity.
Negative points	The combustion of fossil fuels produces carbon dioxide; this is linked to global warming.	Harmful radiation is given out when nuclear fuels break down. The products are also radioactive.

Fuels from petroleum

As the table above shows, petroleum is a fossil fuel. We obtain many other fuels from it. For example propane (C_3H_8), butane (C_4H_{10}), and the mixture of compounds we call petrol. All burn in air as methane does, giving carbon dioxide and water on complete combustion.

Make the link to ... refining petroleum, on page 140.

Hydrogen as a fuel

The reaction of hydrogen with oxygen gives out a great deal of energy:

$2H_2 (g) + O_2 (g) \rightarrow 2H_2O (g)$ + energy

In fact a mixture of hydrogen and oxygen explodes when lit.
- Hydrogen is used to fuel rockets, for example to lift satellites into orbit; the rockets carry fuel tanks of hydrogen and oxygen.
- The product of the combustion reaction, water, is harmless.
- In fuel cells, hydrogen reacts safely with oxygen without combustion. (See page 62.)

Quick check for 8.3 *(Answers on page 166)*
1. Give two key differences between fossil fuels and nuclear fuels, as energy sources.
2. *Combustion* means ...? Complete the sentence.
3. Give one advantage of hydrogen, over fossil fuels and nuclear fuels.
4. Nitrogen reacts with oxygen at high temperatures to form nitrogen dioxide:
   $N_2 + 2O_2 \rightarrow 2NO_2$
   The energy change is +68 kJ. What can you say about nitrogen as a fuel?

Extended

8.4 Cells as a source of energy

Cells are also used to provide energy. In a cell, chemical energy is converted into an electric current.

Simple cells

A simple cell consists of rods of two different metals, standing in an electrolyte. Because one metal gives up electrons more readily than the other, a current flows between them. Look at this example:

3 The electrons flow off through the wire as an electric current.

bulb   voltmeter

4 The current will make a bulb light. You can also measure the voltage that 'pushes' it.

magnesium strip ——   —— copper strip

2 Magnesium is more reactive than copper, so it gives up electrons, and goes into solution as ions:
$$Mg\,(s) \longrightarrow Mg^{2+}\,(aq) + 2e^-$$

So magnesium atoms are oxidised. The magnesium strip loses mass.

solution of sodium chloride

5 The electrons flow to the copper strip. There they are accepted by H+ ions (since hydrogen is less reactive than sodium):
$$2H^+\,(aq) + 2e^- \longrightarrow H_2\,(g)$$

So hydrogen ions are reduced. The copper strip is unchanged.

1 The electrolyte is a solution of sodium chloride. As well as Na^+ and Cl^- ions, it also contains H^+ and OH^- ions from water. (See page 51.)

So the redox reactions provide a current, in a simple cell.

What happens when you change the metals?

Changing the metals also changes the readings on the voltmeter. For example:

Metal strips	Voltage (volts)
magnesium and copper	2.7
iron and copper	0.78
iron and lead	0.32

The further apart the two metals are in the reactivity series, the higher the voltage of the cell.
The more reactive metal in each pair gives up electrons.

Reminder: the reactivity series

potassium
sodium
calcium
magnesium
aluminium   increasing
iron   reactivity
lead
copper
silver
gold

Batteries

Batteries such as torch batteries are just simple cells, in portable form.

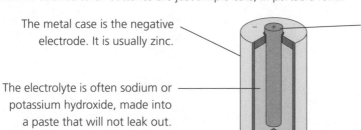

The metal case is the negative electrode. It is usually zinc.

The electrolyte is often sodium or potassium hydroxide, made into a paste that will not leak out.

The positive electrode is in the centre. It is often manganese(IV) oxide packed around a carbon rod.
(Mn^{4+} ions accept electrons, to become Mn^{3+} ions.)

The battery 'dies' when the reactions at the electrodes stop.

Fuel cells

In a fuel cell, hydrogen reacts with oxygen without combustion.
Energy is produced in the form of an electric current.

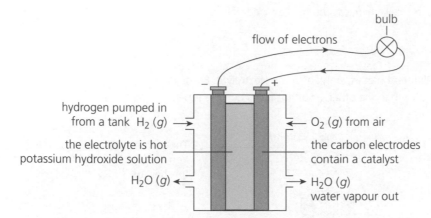

At the negative pole (−)	At the positive pole (+)
• Hydrogen reacts with OH⁻ ions, losing electrons: $2H_2 (g) + 4OH^- (aq) \longrightarrow 4H_2O (g) + 4e^-$ So the hydrogen is **oxidised**.	• Oxygen gains electrons: $O_2 (g) + H_2O (g) + 4e^- \longrightarrow 4OH^- (aq)$ So the oxygen is **reduced**.
• The electrons flow along the wire to the positive pole, as a current. On the way the current can be used to light bulbs, power a car, and so on.	• The OH⁻ ions flow through the electrolyte to the negative pole.

So the redox reactions provide a current, in the fuel cell.

The overall reaction is: $2H_2 (g) + O_2 (g) \longrightarrow 2H_2O (g)$

Uses of fuel cells

- They are used in spacecraft, to provide electricity.
- Since they produce only water, and no carbon dioxide or harmful radiation, there is growing interest in fuel cells as a clean form of energy.
- They are already being used in some homes, to give electricity.
- They are being developed for cars, motorbikes and buses (instead of petrol engines).

✓
Quick check for 8.4   (Answers on page 166)

1 What does a simple cell consist of?
2 Which way does the current flow, in a simple cell?
3 A simple cell was set up, consisting of strips of copper and zinc immersed in sodium chloride solution.
   a Which metal becomes oxidised? Write the half equation for this.
   b What is the function of the sodium chloride solution?
4 a Using the table on page 61, predict the voltage of a lead – copper cell.
   b Which of the two metals goes into solution as ions, in this cell?
5 What is the overall reaction, in a hydrogen fuel cell?
6 *Hydrogen fuel cells do not contribute to global warming.* Explain this statement.

Questions on section 8

Answers for these questions are on page 166.

Core curriculum

1 The diagram on the right represents the energy change when pure sulfuric acid, H_2SO_4 (*l*), is added to water. Choose the correct statements for this reaction.

i   A   The temperature of the solution will rise.

   or B   The temperature of the solution will fall.

ii   A   The products have lower energy than the reactants.

   or B   The products have higher energy than the reactants.

iii A   Heat is taken in by the chemicals.

   or B   Heat is given out by the chemicals.

iv   A   It is an endothermic reaction.

   or B   It is an exothermic reaction.

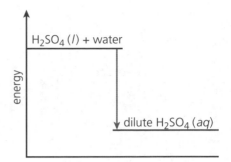

2 Hydrogen can be obtained from water by electrolysis. It can be used as a fuel.

a Write the equation for the burning of hydrogen.

b Suggest why hydrogen is a renewable source of energy.

c When hydrogen is burnt, heat is given off. Which type of reaction gives off heat?

CIE 0620 June '07 Paper 2 Q3

Extended curriculum

Extended

1 Ammonia is made from nitrogen and hydrogen: N_2 (*g*) + $3H_2$ (*g*) $\longrightarrow$ $2NH_3$ (*g*)

a Complete this table for the breaking and making of bonds, in the reaction:

Bonds	Energy change / kJ	Exothermic or endothermic
1 mole of N≡broken	+945	**i**...
ii.......... moles of H—H broken	+1308	**iii**...
6 moles of **iv**............... formed	−2328	**iv** ....

b Use the data above to decide whether the formation of ammonia from nitrogen and hydrogen is exothermic or endothermic.

2 The alcohols form a homologous series. The first four members are methanol, ethanol, propan-1-ol and butan-1-ol.

a One characteristic of a homologous series is that the physical properties vary in a predictable way. The table below gives the heats of combustion of three alcohols.

Alcohol	Formula	Heat of combustion in kJ/mol
methanol	CH_3OH	−730
ethanol	$CH_3{-}CH_2{-}OH$	−1370
propan-1-ol	$CH_3{-}CH_2{-}CH_2{-}OH$	−2020
butan-1-ol	$CH_3{-}CH_2{-}CH_2{-}CH_2{-}OH$	

i   The minus sign indicates that there is less chemical energy in the products than in the reactants. What form of energy is given out by the reaction?

ii   Is the reaction exothermic or endothermic?

iii Write the equation for the complete combustion of ethanol.

iv   Draw an energy level diagram for this reaction.

b Predict a value for the heat of combustion of butan-1-ol.

Explain how you reached this value.   *CIE 0620 November '07 Paper 3 Q6*

3 In this list of ionic equations, the metals are in order of reactivity, the most reactive first.

$Zn \longrightarrow Zn^{2+} + 2e^-$

$Sn \longrightarrow Sn^{2+} + 2e^-$

$Ag \longrightarrow Ag^+ + e^-$

a The following diagram shows a simple cell.

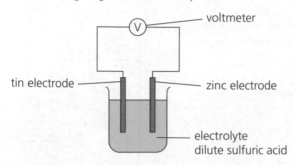

voltmeter

tin electrode —

zinc electrode

electrolyte
dilute sulfuric acid

  i Predict how the voltage of the cell would change if the tin electrode was
  replaced with a silver one.

  ii Which electrode would go into solution as positive ions?
  Give a reason for your choice.

  iii State how you can predict the direction of the electron flow in cells of this type.

b Electrolysis and cells both involve chemical reactions and electricity.
What is the essential difference between them?

CIE 0620 November '04 Paper 3 Q4

4 **a** Exothermic reactions produce heat energy. An important fuel is methane, natural
gas. The equation for its combustion is as follows:

$CH_4 + 2O_2 \longrightarrow CO_2 + 2H_2O$

In chemical reactions bonds are broken and new bonds are formed.
Using this reaction give an example of:

  i a bond that is broken   **ii** a bond that is formed.

  iii Explain, using the idea of bonds forming and breaking, why this reaction
  is exothermic, that is, it produces heat energy.

b Some radioactive isotopes are used as nuclear fuels.

  i Give the symbol and the nucleon number of an isotope used as a nuclear fuel.

  ii Give another use of radioactive isotopes.

CIE 0620 June '06 Paper 3 Q6

Alternative to practical

1 Diesel is a liquid fuel obtained from crude oil. Biodiesel is a fuel made from oil obtained
from the seeds of plants such as sunflowers. Using the apparatus below, plan an
experiment to investigate which of these two fuels produces more energy.

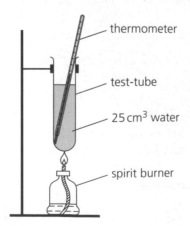

thermometer

test-tube

$25\,cm^3$ water

spirit burner

CIE 0620 November '07 Paper 6 Q7

9 Rates of reaction

The big picture

- The rate of a reaction is a measure of how fast or slowly it goes.
- We can make reactions go faster or more slowly by changing the reaction conditions – for example by raising the temperature.
- The collision theory helps us to explain *why* rates change, as we change reaction conditions.

9.1 What does rate mean?

- **Rate** is a measure of how **fast** a reaction goes – its **speed**.
- To measure the rate of a reaction, you need to do an **experiment**, in which you measure a **change** (for example in mass, or volume) over a period of **time**.

The reaction of magnesium with hydrochloric acid
The equation is:

$$Mg\ (s)\ +\ 2HCl\ (aq)\ \longrightarrow\ MgCl_2\ (aq)\ +\ H_2\ (g)$$

One product is a gas. So you can measure the **change in volume** of the gas, over time.

The apparatus for the experiment
The gas can be collected in a gas syringe, like this:

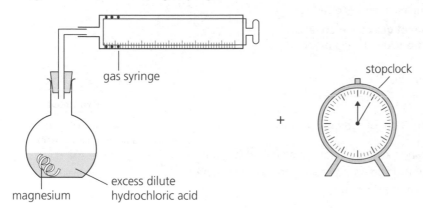

gas syringe

stopclock

+

magnesium

excess dilute hydrochloric acid

or in an upturned measuring cylinder full of water, like this:

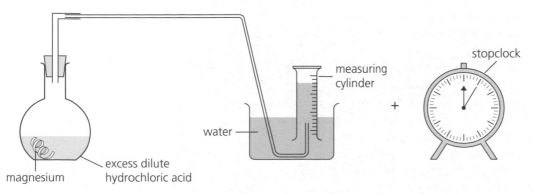

measuring cylinder

stopclock

water

+

magnesium

excess dilute hydrochloric acid

The volume of gas is recorded at regular intervals. The data obtained is used to plot a graph.

A graph of the results

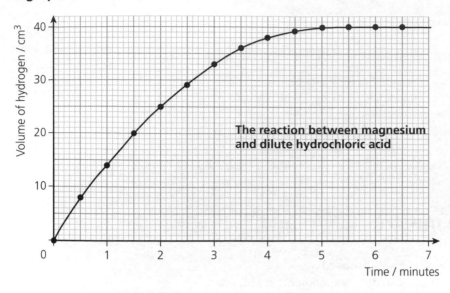

The reaction between magnesium and dilute hydrochloric acid

Note the curved shape of the graph.
*   The slope of the curve tells you how fast the reaction is going.
*   The slope keeps changing until the curve flattens out – which tells us that the rate of the reaction keeps changing.
*   The curve is steepest at the start, so the reaction is fastest then. (Its rate is greatest.)
*   The curve becomes less steep as time passes, showing that the reaction is slowing down.
*   When the curve goes flat, it means the reaction has stopped. The reaction rate is zero.

Working out the average rate of reaction

The *average* rate of the reaction, from beginning to end, is given by:

average rate of reaction
(in cm³ per min or sec) $= \dfrac{\textbf{total volume of gas collected (in cm}^3\textbf{)}}{\textbf{total time taken (in min or sec)}}$

The reaction above lasted for 5 minutes. After that the curve was flat.
40 cm³ of hydrogen was produced.

So average rate of reaction $= \dfrac{40 \text{ cm}^3}{5 \text{ minutes}} = 8$ cm³ of hydrogen/minute.

But remember that this is the average rate, and the rate is in fact continually changing.
For example the graph shows that 14 cm³ of hydrogen was produced in the first minute.
So the rate for that minute was 14 cm³ of hydrogen/minute.

> **Remember**
> When you find rate from a graph, always give the units of time (for example, per minute).

✓ **Quick check for 9.1** *(Answers on page 166)*

1   What is meant by the *rate* of a chemical reaction?
2   What kind of thing do you need to measure, to find the rate of a chemical reaction?
3   Describe how the rate of the reaction changes over time, in the reaction between magnesium and hydrochloric acid.
4   Look at the graph above. How can you tell from it when the reaction was over?
5   In a similar experiment, 60 cm³ of hydrogen was produced in 4 minutes. What was the average rate of the reaction this time?
6   When marble chips react with hydrochloric acid, carbon dioxide gas bubbles off. So there is a loss of mass. (Carbon dioxide is quite a heavy gas.) Suggest a method for finding the rate of this reaction.

9.2 Ways to change the rate of reaction

By changing temperature, concentration, or particle size

You can change the rate of the reaction by changing the temperature, or the concentration of a reactant, or the particle size of a solid reactant. Look at this table:

When you increase ...	by ...	the rate ...	
the temperature	heating	increases	↑
the concentration of a reactant	using a more concentrated solution	increases	↑
the surface area of a solid reactant	using smaller pieces (which means a larger total surface area)	increases	↑

When you decrease ...	by ...	the rate ...	
the temperature	cooling	decreases	↓
the concentration of a reactant	using a more dilute solution	decreases	↓
the surface area of a solid reactant	using larger pieces (which means a smaller total surface area)	decreases	↓

But note that the amount of product obtained in the reaction does not change. It stays the same, no matter how the rate changes.

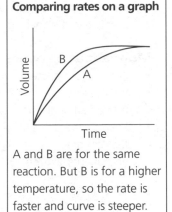

Comparing rates on a graph

A and B are for the same reaction. But B is for a higher temperature, so the rate is faster and curve is steeper.

Some explosive reactions

Some reactions go so fast under some conditions that you get an explosion!
- In flour mills, the air can fill with fine flour dust, with a very large total surface area. A spark can cause the flour to catch fire and explode. (It reacts with the oxygen in air.)
- In coal mines, methane and other flammable gases can collect in the air. At certain concentrations they form an explosive mix, which can be ignited by a spark.

By using a catalyst

A **catalyst** is a substance that speeds up a chemical reaction, without being used up itself.

Example: decomposition of hydrogen peroxide

Hydrogen peroxide decomposes, giving off oxygen. The equation is:

$2H_2O_2 (aq) \longrightarrow 2H_2O (l) + O_2 (g)$

Now look how a catalyst affects the rate of reaction:

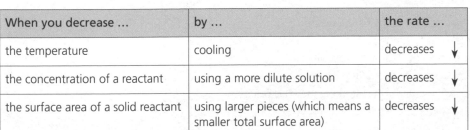

1 Under normal conditions the reaction is very slow.	**2** Add a little manganese(IV) oxide.	**3** Or a little raw liver, which contains the **enzyme** catalase.
Even if you put a stopper on the tube and leave it for hours, there there is not enough oxygen to relight the glowing splint. (That is the test for oxygen.)	The glowing splint relights straight away. So the reaction has speeded up. Manganese(IV) oxide is a catalyst for this reaction. (But is not itself changed.)	Once again there is enough oxygen to relight the glowing splint. So the enzyme catalase also acts as a catalyst for this reaction.

Catalysts in industry

- Catalysts speed up reactions, so they save time. They can also reduce fuel bills, if the reaction is fast enough at a lower temperature. So they are very important in industry.
- The catalysts used are usually transition metals, or oxides of transition metals. For example iron in making ammonia, and vanadium pentoxide in making sulfuric acid.

> **Remember**
> A catalyst changes the rate of a reaction, but NOT the amount of product obtained.

More about enzymes

- Enzymes are **proteins** made by living cells. They act as **biological catalysts**.
- We use enzymes for many purposes. The enzymes in yeast cells are used as a catalyst in making ethanol. Enzymes made by bacteria are used in biological detergents.
- Enzymes work in a limited range of temperature and pH, similar to the conditions in the cells that made them. If the temperature is too high, or the pH too different, they become **denatured** and stop working.

✓

Quick check for 9.2   (*Answers on page 166*)

1 How does the rate of a chemical reaction change when:
   a the temperature is increased?   **b** the particle size of a solid is increased?
2 Which has a larger surface area: 2g of powdered chalk, or a 2g lump of chalk?
3 Give: **a** one similarity   **b** one difference
   between the enzyme catalase, and the catalyst manganese(IV) oxide.

9.3 The collision theory

Extended

The collision theory is a way to explain reaction rates. These are its key points:
- For a reaction to take place, there must be a **collision** between the reactant particles.
- But not every collision leads to reaction. Some are **unsuccessful**.
- A **successful** collision is one where the colliding particles have enough energy to react.

These diagrams represent the reaction of magnesium with a solution of hydrochloric acid

① 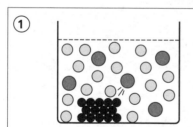	② 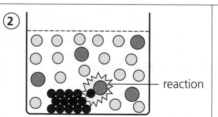	● = add particle ● = magnesium atom ○ = water particle
This was an unsuccessful collision: not enough energy for a reaction.	This is a successful collision – the particles have enough energy to react.	

③ 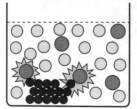	④ 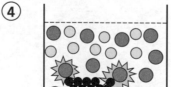	⑤ 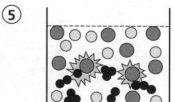
If the **temperature** is increased, the particles have more energy. They move faster, and collide more often – and more of the collisions are successful.	If the **concentration** of the acid is increased, it means there are more acid particles in the same volume. So there are more successful collisions.	Here the **surface area** of the metal is increased, by using magnesium powder. So more magnesium atoms are exposed, giving more successful collisions.

The more successful collisions there are, the faster the reaction goes.

The collision theory and gas reactions

In reactions between gases, the rate increases as temperature and pressure increase.

- As the **temperature** rises, the gas molecules gain more energy and move faster.
- Increasing the **pressure** pushes the gas molecules closer together.

So in each case, the chance of successful collisions increases.

Quick check for 9.3   *(Answers on page 166)*

1   What does the collision theory say?
2   Use the theory to explain why *lowering* the temperature slows down a reaction.
3   When two gases react, raising the pressure increases the reaction rate. Why?

9.4 Reactions that need light

Photochemical reactions take place only in the presence of light. Here are two examples.

Example 1	Photosynthesis
Where it takes place	In the leaves of plants.
The photochemical reaction	Carbon dioxide from the air, and water from the soil, react to produce **glucose**, using energy from sunlight.
Equation	$6CO_2$ (g) + $6H_2O$ (l) $\longrightarrow$ $C_6H_{12}O_6$ (s) + $6O_2$ (g)
Catalyst	Chlorophyll – the green pigment in the leaves.
Effect of increasing the light intensity	The reaction speeds up.
Application	Plants use the glucose for their food – and we use the oxygen they give off, for respiration in our body cells. (We breathe it in.)

Example 2	Reactions on photographic film
Where it takes place	On photographic film, which is coated with a gel containing silver halides.
The photochemical reaction	When the camera shutter is opened and light strikes the film, the silver ion in the silver halides is **reduced** to silver.
Equation	Ag^+ (s) + e^- $\longrightarrow$ Ag (s)
Catalyst	None.
Effect of increasing the light intensity	The reaction speeds up. Silver forms fastest where the brightest light strikes the film.
Application	The deposit of silver forms dark areas on the black-and-white developed film. The darkest areas correspond to the brightest parts of the scene. Later, light is shone through the film onto special photographic paper, which is coated with silver halides. So the same photochemical reaction takes place. The darkest areas on the film let least light through, so appear brightest on the paper. The result is a copy of the scene, in shades from black to white.

Quick check for 9.4   *(Answers on page 166)*

1   What is meant by a *photochemical* reaction?
2   Write a word equation for the photosynthesis reaction.
3   Explain why different shades of grey can be obtained on a photo, in black-and-white photography using film.

Extended

Questions on section 9

Answers for these questions are on page 166.

Core curriculum

1 Lumps of calcium carbonate react with hydrochloric acid, releasing carbon dioxide gas:

$$CaCO_3\,(s) + 2HCl\,(aq) \longrightarrow CaCl_2\,(aq) + CO_2\,(g) + H_2O\,(l)$$

   a Describe a practical method for investigating this reaction, which would enable you to calculate the rate of reaction.

   b What effect will the following have on the rate of the reaction?

      i increasing the temperature   **ii** adding water to the acid

      iii using powdered calcium carbonate instead of lumps.

CIE 0620 June '04 Paper 2 Q3

2 Catalysts are often used in industry.

   a **i** What do you understand by the term catalyst?

      ii Which type of metals often act as catalysts?

   b A student measured the volume of hydrogen gas produced when a few large pieces of zinc reacted with hydrochloric acid of concentration 2.0 mol/dm³. The hydrochloric acid was in excess. The results are given in the table.

Time/minutes	0	10	20	30	40	50	60
Volume of hydrogen/cm³	0	27	54	81	100	110	110

      i Plot a graph of volume of hydrogen against time. Label the axes.

      ii Copper ions catalyse this reaction. On the same axes, sketch the line you would expect for the catalysed reaction. Label this line C.

      iii Explain why no more hydrogen is given off after 50 minutes.

   c What would happen to the speed of the reaction if:

      i small pieces of zinc were used instead of large pieces,

      ii the concentration of hydrochloric acid was 1.0 mol/dm³?

CIE 0620 June '07 Paper 2 Q4

Extended curriculum

1 **a** A small piece of marble, calcium carbonate, was added to 5 cm³ of hydrochloric acid at 25°C. The time taken for the reaction to stop was measured.

$$CaCO_3\,(s) + 2HCl\,(aq) \longrightarrow CaCl_2\,(aq) + CO_2\,(g) + H_2O\,(l)$$

Similar experiments were performed, always using 5 cm³ of hydrochloric acid.

Experiment	Number of pieces of marble	Concentration of acid mol/dm³	Temperature/°C	Time/min
1	1	1.00	25	3
2	1	0.50	25	7
3	1 piece crushed	1.00	25	1
4	1	1.00	35	2

Explain each of the following in terms of collisions between reacting particles.

      i Why is the rate in experiment 2 slower than in experiment 1?

      ii Why is the rate in experiment 3 faster than in experiment 1?

      iii Why is the rate in experiment 4 faster than in experiment 1?

   b An alternative method of measuring the rate of this reaction would be to measure the volume of carbon dioxide produced at regular intervals.

      i Sketch a graph showing the volume of carbon dioxide against time.

      ii Show how the rate of reaction could be obtained from the graph.

CIE 0620 November 07 Paper 3 Q7

Extended

Extended

2 Three of the factors that can influence the rate of a chemical reaction are:
- physical state of the reactants
- light
- the presence of a catalyst

a The first recorded dust explosion was in a flour mill in Italy in 1785. Flour contains carbohydrates. Explosions are very fast exothermic reactions.

   i Use the collision theory to explain why the reaction between the particles of flour and the oxygen in the air is very fast.

   ii Write a word equation for this exothermic reaction.

The decomposition of silver(I) bromide is the basis of film photography. This reaction is photochemical. The equation for this decomposition is:

$$2AgBr \longrightarrow 2Ag + Br_2$$
   white   black

A piece of white paper was coated with silver(I) bromide and the following experiment was carried out.

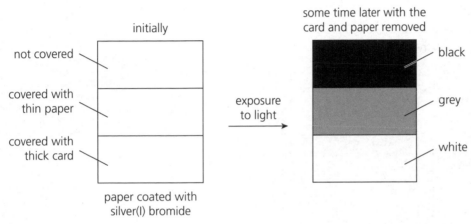

paper coated with
silver(I) bromide

b Explain the results.   *CIE 0620 June '08 Paper 3 Q6*

3 The rate of a reaction depends on concentration of reactants, temperature and possibly a catalyst or light.

a A piece of magnesium ribbon was added to 100 cm³ of 1.0 mol/dm³ hydrochloric acid. The hydrogen evolved was collected in a gas syringe and its volume measured every 30 seconds. The **acid** was in **excess**.

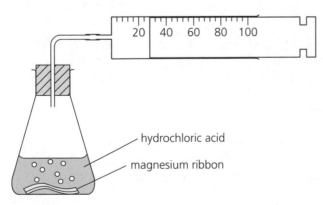

hydrochloric acid

magnesium ribbon

The results are given in the table.

Time/seconds	0	30	60	90	120	150	180	210	240	270	300	330	360
Volume of hydrogen/cm³	0	15	25	33	41	47	52	56	58	59	60	60	60

   i Plot a graph of volume of hydrogen against time. Label the axes.

   ii   The experiment was repeated. Two pieces of magnesium ribbon were added to 100 cm^3 of 1.0 mol/dm^3 hydrochloric acid. Sketch this graph on the same grid as your graph for **i**, and label it X.

   iii The experiment was repeated using one piece of magnesium ribbon and 100 cm^3 of 2.0 mol/dm^3 hydrochloric acid. Describe how the **shape** of this graph would differ from the one you plotted.

b   Reaction rate increases when concentration or temperature is increased. Using the idea of reacting particles, explain why;

   i   increasing concentration increases reaction rate,

   ii increasing temperature increases reaction rate.

c   The rate of a photochemical reaction is affected by light. A reaction, in plants, between carbon dioxide and water is photochemical.

   i   Name the two products of this reaction.

   ii This reaction will only occur in the presence of light and another chemical. Name this chemical.   *CIE 0620 November '06 Paper 3 Q7*

Alternative to practical

1   Magnesium reacts with dilute sulfuric acid to form hydrogen gas. The speed of the reaction was investigated using the apparatus below.

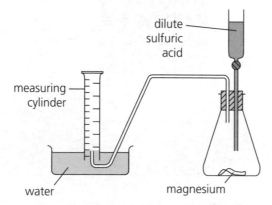

In an experiment 50 cm^3 of dilute sulfuric acid was added to a large piece of magnesium. A student measured the total volume of gas produced at 2 minute intervals. Here are the results.

Time / minutes	0	2	4	6	8	10	12
Water level in measuring cylinder	0 / 5 / 10	10 / 15 / 20	25 / 30 / 35	25 / 30 / 35	35 / 40 / 45	40 / 45 / 50	40 / 45 / 50

a   Plot a graph of volume against time. Label the axes. Join the points with a smooth curve.

b **i**   At which time does the result appear to be inaccurate?

     ii Use the graph to deduce what the correct volume should be at this time.

          CIE 0620 November '07 Paper 6 Q6

2   **Is manganese(IV) oxide a catalyst?**

A catalyst speeds up a chemical reaction, and remains unchanged. Hydrogen peroxide, H_2O_2, breaks down to form oxygen. This reaction is very slow without a catalyst.

Describe an experiment to show that manganese(IV) oxide is a catalyst for this reaction You are provided with the following items.

hydrogen peroxide solution, manganese(IV) oxide, distilled water, measuring cylinder, balance, beaker, filtration apparatus, splints, Bunsen burner.

          CIE 0620 June '04 Paper 6 Q4

10 Reversible reactions, and equilibrium

The big picture
- Some reactions are reversible: they can go backwards as well as forwards.
- In a closed container, these reactions do not complete. They reach a state of equilibrium, with a mixture of reactants and products present.
- But we usually want to obtain as much product as possible.
- There are steps we can take to change the equilibrium, and obtain more product.

10.1 What are reversible reactions?

Many reactions are **reversible** – they can go back the way they came!

Reversible reactions	Equations
- produce products, like all reactions do	reactants $\longrightarrow$ products We call this step the **forward** reaction.
- go backwards too	reactants $\longleftarrow$ products We call this step the **backward** reaction.
- may have both forward and backward reactions going on at the same time, in the container	reactants $\longrightarrow$ products *and* reactants $\longleftarrow$ products
- so are shown by a special symbol	reactants $\rightleftharpoons$ products

Example 1: two forms of copper(II) sulfate

	Hydrated copper(II) sulfate	Anhydrous copper(II) sulfate
Chemical formula	$CuSO_4 . 5H_2O$	$CuSO_4$
Appearance	blue solid	white solid
When you heat hydrated copper(II) sulfate, the colour changes from blue to white, and steam comes off.	$CuSO_4 . 5H_2O \ (s) \longrightarrow CuSO_4 \ (s) + 5H_2O \ (l)$ This is the forward reaction. It is an **endothermic** reaction: you must put heat energy in.	
If you add water to anhydrous copper(II) sulfate, the colour changes from white to blue, and the solid gets hot.	$CuSO_4 . 5H_2O \ (s) \longleftarrow CuSO_4 \ (s) + 5H_2O \ (l)$ This is the backward reaction. It is an **exothermic** reaction: energy is given out.	
The equation uses the special symbol, to show that it is a reversible reaction.	$CuSO_4 . 5H_2O \ (s) \rightleftharpoons CuSO_4 \ (s) + 5H_2O \ (l)$ This is the overall equation for the reaction.	

Note
- Hydrated copper(II) sulfate is also called copper(II) sulfate-5-water.
- The water in copper(II) sulfate-5-water is called **water of crystallisation**.

A test for water
Since white anhydrous copper(II) sulfate turns blue when water is added, it can be used to test for water.
- Add a few drops of the unknown liquid to anhydrous copper(II) sulfate.
- If the compound turns blue, there is water in the liquid.
- If the liquid also boils at 100 °C, and freezes at 0 °C, it is pure water.

Example 2: the reaction between nitrogen and hydrogen

Nitrogen and hydrogen react together to form ammonia – but the reaction is reversible:

Forward reaction	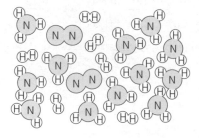	$N_2 (g) + 3H_2 (g) \longrightarrow 2NH_3 (g)$ Exothermic reaction: energy is given out.
Backward reaction		$2NH_3 (g) \longleftarrow N_2 (g) + 3H_2 (g)$ Endothermic reaction: energy is taken in.
Overall equation		$N_2 (g) + 3H_2 (g) \rightleftharpoons 2NH_3 (g)$

> **Remember**
>
> If the forward reaction is exothermic, the backward reaction is endothermic ... and vice versa.

✓ **Quick check for 10.1** *(Answers on page 167)*

1 What is a *reversible* reaction?
2 How can you tell from a chemical equation whether a reaction is reversible or not?
3 What would you observe, if you heated copper(II) sulfate-5-water gently?
4 How would you reverse the reaction you described in **3**?
5 Hydrated cobalt(II) chloride has the formula $CoCl_2.6H_2O$. It reacts similarly to hydrated copper(II) sulfate when it is heated gently, and the reaction is reversible. Write the overall equation for it.

10.2 Reversible reactions and equilibrium

What is equilibrium?

In a closed system (where no reactants or products can escape) a reversible reaction will reach a state of **dynamic equilibrium**, where:

* the forward and backward reactions are taking place **at exactly the same rate**
* so the amounts of reactants and products present do not change.

For example, look at the reaction for making ammonia: $N_2 (g) + 3H_2 (g) \rightleftharpoons 2NH_3 (g)$

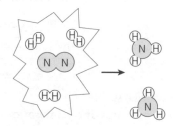
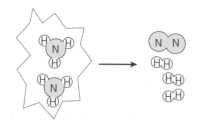

The reaction has reached dynamic equilibrium. So the amount of ammonia present will not increase ...

... because every time nitrogen and hydrogen molecules react to give ammonia molecules ...

... other ammonia molecules break down to give nitrogen and hydrogen molecules.

The word **dynamic** means that reactions are still going on, all the time, in the container. We usually shorten **dynamic equilibrium** to just **equilibrium**.

Shifting the equilibrium

Here again is the equation for making ammonia:

$$N_2 (g) + 3H_2 (g) \rightleftharpoons 2NH_3 (g)$$

By changing conditions, we can **shift the equilibrium** for the reaction. (*Shift* means *move*.)

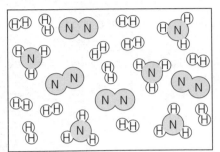

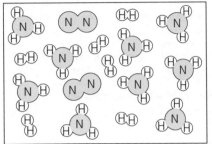

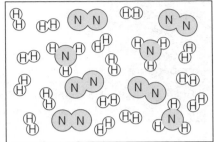

Look at this equilibrium mixture, for the reaction. As you would expect, it contains nitrogen, hydrogen, and ammonia.

By changing conditions, we can obtain more ammonia. We have shifted the equilibrium to the right (or towards the product, in the equation).

Or with different conditions, we can reduce the amount of ammonia. We have shifted the equilibrium to the left (towards the reactants, in the equation).

Conditions that can be changed

This table shows conditions that can be changed, for reversible reactions, and the result.

The change	How the equilibrium shifts
Increase the pressure (for reactions involving gases)	Equilibrium shifts to favour the side of the equation with **fewer gas molecules** (that is, with a lower volume of gas).
Reduce the pressure (for reactions involving gases)	Equilibrium shifts to favour the side of the equation with **more gas molecules** (a higher volume of gas).
Increase the temperature (any reaction)	Equilibrium shifts to favour the **endothermic** reaction.
Reduce the temperature (any reaction)	Equilibrium shifts to favour the **exothermic** reaction.
Add a catalyst (any reaction)	Both the forward and back reactions speed up – and by the same amount. So there is **no shift** in the equilibrium, but the reaction reaches equilibrium faster.

Note
Increasing the temperature shifts equilibrium – AND speeds up both the forward and backward reactions.

Why does equilibrium shift?

When a reaction has reached equilibrium, and you change the temperature or pressure, *it acts to oppose the change.*

For example, in the reaction for making ammonia:

$$N_2 (g) + 3H_2 (g) \underset{\text{endothermic}}{\overset{\text{exothermic}}{\rightleftarrows}} 2NH_3 (g)$$

4 molecules   2 molecules

Remember
In making ammonia from nitrogen and hydrogen, the forward reaction is exothermic.

- If you increase the pressure, you in effect push more molecules into a smaller space. So more ammonia forms, to reduce the number of molecules in that space – and therefore reduce the pressure. (Four molecules become two.)
- But if you increase the temperature, more ammonia breaks down, to absorb the extra heat energy you have added. (The back reaction is endothermic.)

The reaction mixture then reaches a new state of equilibrium, with different proportions of nitrogen, hydrogen, and ammonia in the mix.

Shifting the equlibrium for ammonia: a summary

This summarises how changing the conditions affects the amount of ammonia obtained:

The reactions	$N_2(g) + 3H_2(g) \xrightleftharpoons[\text{endothermic}]{\text{exothermic}} 2NH_3(g)$ 4 molecules 2 molecules
Increase the pressure	Increasing the pressure will give **more** ammonia, since there are fewer molecules on the right. Equilibrium shifts to the right.
Reduce the temperature	Reducing the temperature will give **more** ammonia, since the forward reaction is exothermic. Equilibrium shifts to the right.
Add a catalyst (iron)	**No change** in the amount of ammonia obtained. The catalyst does not shift equilibrium. But it allows the reaction to reach equilibrium faster.

The chosen conditions for making ammonia

Ammonia is a very important compound. (For example, it is used in making nitric acid, and fertilisers.) So we want a high yield: we want to shift the equilibrium as far to the right as we can. The table above shows that the best conditions for manufacturing ammonia are:

- high pressure
- low temperature
- iron catalyst.

But the lower the temperature, the slower the reaction. A slow reaction is not desirable in industry, since time is money. So a compromise temperature is chosen.

> **Make the link to ...**
> the Haber process for making ammonia, on page 133.

Changing concentration in equilibrium reactions

Equilibrium can also be shifted in reversible reactions involving solutions:

reactants $\rightleftharpoons$ products

You can shift the equilibrium by changing the concentration of a reactant or product in the reaction mixture. Each time, the system acts to oppose the change:

Change made to the mixture	Result
Add more reactant	Equilibrium shifts to the right. More product is produced.
Remove some reactant	Equilibrium shifts to the left. More product breaks down to form reactants.
Add more product	Equilibrium shifts to the left. More product breaks down to form reactants.
Remove some product	Equilibrium shifts to the right. More product is produced.

For example, iodine reacts reversibly with dilute sodium hydroxide solution:

$I_2(aq) + 2OH^-(aq) \rightleftharpoons I^-(aq) + IO^-(aq) + H_2O(l)$

red-brown   colourless   colourless

If you add more sodium hydroxide, the equilibrium moves to the right, and the solution goes colourless. But if you add dilute acid, its hydrogen ions react with hydroxide ions to form water. So the equilibrium shifts to the left, and the solution turns red-brown.

✓
Quick check for 10.2   *(Answers on page 167)*

1. A reaction reached *equilibrium*. Explain what this term means.
2. Give three ways in which equilibrium can be shifted.
3. In a reversible reaction, raising the temperature lowered the yield of product. What does this tell you about the reaction?
4. Why does a catalyst *not* shift equilibrium?
5. In the reaction $2SO_2(g) + O_2(g) \rightleftharpoons 2SO_3(g)$, the forward reaction is exothermic. How would you increase the yield of the product?

Questions on section 10

Answers for these questions are on page 167.

Extended

Extended curriculum

1   Ammonia is made from nitrogen and hydrogen. The energy change in the reaction
    is − 92 kJ/mole. The reaction is reversible, and reaches equilibrium.
    a   Write the equation for the reaction.
    b   Is the forward reaction endothermic, or exothermic? How can you tell?
    c   Explain why the yield of ammonia:
        i   rises if you increase the pressure
        ii   falls if you increase the temperature
    d   What effect does increasing: i   the pressure   ii   the temperature
        have on the rate at which ammonia is made?
    e   Why is the reaction carried out at 450°C rather than at a lower temperature?

2   Carbonyl chloride, $COCl_2$, is a colourless gas. It is made by the following reaction.

$$CO\ (g) + Cl_2\ (g) \underset{heat}{\overset{cool}{\rightleftharpoons}} COCl_2\ (g)$$

    a   When the pressure on the equilibrium mixture is decreased, the position of
        equilibrium moves to the left.
        i   How does the concentration of each of the three chemicals change?
        ii   Explain why the position of equilibrium moves to the left.
    b   Using the information given with the equation, is the forward reaction exothermic
        or endothermic? Give a reason for your choice.   *CIE 0620 June '08 Paper 3 Q5*

3   The dichromate and chromate ions, $Cr_2O_7^{2-}$ and CrO_4^{2-}, exist in equilibrium like this:
    $Cr_2O_7^{2-}\ (aq) + H_2O\ (l) \rightleftharpoons 2CrO_4^{2-}\ (aq) + 2H^+(aq)$
       orange   yellow
    a   What would you see if you added dilute acid to a solution containing
        chromate ions?
    b   How would you reverse the change?
    c   Explain why adding hydroxide ions shifts the equilibrium.

4   Hydrogen and bromine react reversibly:
    $H_2\ (g) + Br_2\ (g) \rightleftharpoons 2HBr\ (g)$
    a   Which of these will favour the formation of more hydrogen bromide?
        i   adding more hydrogen
        ii   removing bromine
        iii removing the hydrogen bromide as it forms.
    b   Increasing the pressure will have no effect on the amount of product. Explain why.
    c   However, the pressure is likely to be increased, when the above reaction
        is carried out in industry. Suggest a reason for this.

Alternative to practical

1   Hydrated copper sulfate crystals, $CuSO_4.5H_2O$ were heated in this apparatus.
    a   Copy the diagram and indicate using arrows:
        i   where the copper sulfate crystals are placed
        ii   where heat is applied.
    b   What is the purpose of the ice?
    c   Describe the colour change in the crystals.
    d   Is it possible to reverse this change? Explain your answer.

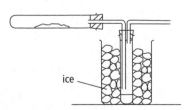

CIE 0620 November '07 Paper 6 Q1

11 Redox reactions

The big picture

- Many reactions in chemistry are redox reactions.
- The word redox is short for *reduction and oxidation*.
- In their simplest definitions, reduction means loss of oxygen, and oxidation means gain of oxygen.
- In their broader definitions, reduction means gain of electrons, and oxidation means loss of electrons.

11.1 Gain and loss of oxygen

The early chemists studied many reactions that involved oxygen. They used these definitions:

- **Oxidation means oxygen is gained.**
- **Reduction means oxygen is lost.**

Example: the reaction between copper(II) oxide and hydrogen

When you pass hydrogen over heated copper(II) oxide, the black powder turns red-brown, because it is being converted to copper:

black copper(II) oxide

hydrogen in

heat

The reaction is shown by this equation:

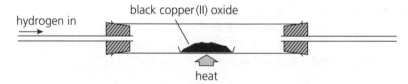

reduction –
copper loses oxygen

$$CuO\ (s) + H_2\ (g) \longrightarrow Cu\ (s) + H_2O\ (l)$$

oxidation –
hydrogen gains oxygen

So copper is reduced, and hydrogen is oxidised, in this reaction.
Reduction and oxidation always take place together, in a reaction.
So the reaction is called a **redox** reaction.

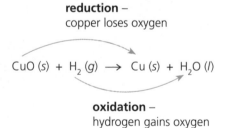

Quick check for 11.1   *(Answers on page 167)*

1. What would you see as hydrogen is passed over hot copper(II) oxide?
2. Using the equation, show that copper(II) oxide is *reduced* in the reaction in **1**.
3. $2Mg\ (s) + CO_2\ (g) \longrightarrow 2MgO\ (s) + C\ (s)$
   In this reaction, which substance is: **a** oxidised?   **b** reduced?
4. What is *redox* short for?
5. Look at this reaction: $CO_2\ (g) + C\ (s) \longrightarrow 2CO\ (g)$.
   Is it a redox reaction? Explain your answer.

Extended

11.2 Electron transfer

The early chemists did not know about electrons, when they studied gain and loss of oxygen. But now we know that when oxygen is gained or lost, electrons are also being transferred. So we use much broader definitions:

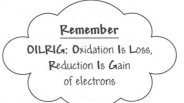

- **In oxidation, electrons are lost.**
- **In reduction, electrons are gained.**

Example 1: magnesium and oxygen

When magnesium burns in oxygen, this reaction takes place:

$2Mg\ (s)\ +\ O_2\ (g)\ \longrightarrow\ 2MgO\ (s)$

Magnesium oxide is an ionic solid, made up of magnesium ions (Mg^{2+}) and oxide ions (O^{2-}). So electrons have been transferred:

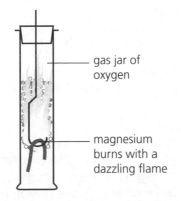

gas jar of
oxygen

magnesium
burns with a
dazzling flame

reduction –
oxygen gains electrons

$2Mg\ +\ O_2\ \longrightarrow\ 2Mg^{2+}O^{2-}$

oxidation –
magnesium loses electrons

So the magnesium has been oxidised, and the oxygen has been reduced.
Reduction and oxidation always take place together, in a reaction.
So the reaction is a redox reaction.

We use two **half equations** to show the electron loss and gain.
For the reaction above, the half equations are:

$Mg\ \longrightarrow\ Mg^{2+}\ +\ 2e^-$ $\qquad\qquad$ $O_2\ +\ 4e^-\ \longrightarrow\ 2O^{2-}$

Example 2: chlorine water and potassium iodide solution

A solution of potassium iodide contains potassium ions (K^+) and iodide ions (I^-).
Chlorine reacts with the iodide ions, giving chloride ions and iodine. The ionic equation is:

$Cl_2\ (aq)\ +\ 2I^-\ (aq)\ \longrightarrow\ 2Cl^-\ (aq)\ +\ I_2\ (aq)$

So electrons have been transferred:

reduction –
chlorine atoms gain electrons

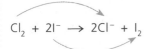

$Cl_2\ +\ 2I^-\ \longrightarrow\ 2Cl^-\ +\ I_2$

oxidation –
iodine ions lose electrons

The reaction is therefore a redox reaction. The half equations are:

$2I^-\ \longrightarrow\ I_2\ +\ 2e^-$ $\qquad\qquad$ $Cl_2\ +\ 2e\ \longrightarrow\ 2Cl^-$

About half equations and ionic equations
- We use **half equations** to show what is happening during a redox reaction. One half equation shows electron loss, the other shows electron gain.
- Half equations are usually balanced, so that both show the same number of electrons.
- Adding the balanced half equations gives the **ionic equation** for the reaction.

From half equations to ionic equations: an example

Chlorine oxidises iron(II) chloride to iron(III) chloride:

$2FeCl_2 \ (aq) + Cl_2 \ (g) \longrightarrow 2FeCl_3 \ (aq)$

So Fe^{2+} ions become Fe^{3+} ions, and chlorine atoms become chloride ions, Cl^-.

The unbalanced half equations	The balanced half equations
$Fe^{2+} \longrightarrow Fe^{3+} + e^-$	$2Fe^{2+} \longrightarrow 2Fe^{3+} + 2e^-$ (multiply by 2)
$Cl_2 + 2e^- \longrightarrow 2Cl^-$	$Cl_2 + 2e^- \longrightarrow 2Cl^-$

Adding the balanced half equations

$$2Fe^{2+} \longrightarrow 2Fe^{3+} + 2e^-$$
$$Cl_2 + 2e^- \longrightarrow 2Cl^-$$

$2Fe^{2+} + Cl_2 + 2e^- \longrightarrow 2Fe^{3+} + 2Cl^- + 2e^-$ 　　　　(the $2e^-$ cancel each other out)

So the **ionic equation** is: $2Fe^{2+} + Cl_2 \longrightarrow 2Fe^{3+} + 2Cl^-$

Adding state symbols gives: $2Fe^{2+} \ (aq) + Cl_2 \ (g) \longrightarrow 2Fe^{3+} \ (aq) + 2Cl^- \ (aq)$

✓

Quick check for 11.2　　　　　　　　　　　　　(*Answers on page 167*)

1　Sodium reacts with chlorine to form sodium chloride.

　　a　Write a balanced equation for the reaction.

　　b　Explain why it is a redox reaction, even though no oxygen is present.

2　a　Which substance is oxidised, and which is reduced, in the reaction in **1**?

　　b　Write balanced half equations for the oxidation and reduction reactions.

11.3 Oxidation state

- Oxidation state is a measure of how oxidised or reduced an element is, in a compound.
- The **oxidation state** is a number, showing how many electrons have been lost, gained, or shared by an element, in forming a compound
- The number is shown using Roman numerals: (0, I, II, III, IV …).

The rules for working out the oxidation state

Rule	Example
1 The oxidation state of an uncombined element is 0.	In H_2, the oxidation state of hydrogen is 0. In O_2, the oxidation state of oxygen is 0.
2 In an ionic compound, the oxidation state of the element is the same as the charge on its ions.	Sodium chloride is made up of Na^+ and Cl^- ions. So the oxidation states are: 　　　　　Na Cl 　　　　　+I −I
3 The oxidation state of hydrogen in most compounds is +I.	H Cl 　　　　　+I −I
4 The oxidation state of oxygen in most compounds is −II.	Mg O 　　　　　+II −II
5 The oxidation states in a compound add up to zero.	In MgO　　Mg = +II　　　In H_2O　　2H = +II (each is +I) 　　　　　　　　O = −II　　　　　　　　　　O = −II 　　　　　　Total　 0　　　　　　　　　　　Total　 0

The more oxidized something is, the higher its oxidation number.
The more reduced it is, the lower its oxidation number.

Extended

Oxidation states and redox reactions

Look at the oxidation states for the reaction between copper(II) oxide and hydrogen:

$$CuO + H_2 \longrightarrow Cu + H_2O$$

+II −II   0   0   2x +I −II

The oxidation state of copper has decreased from +II to 0.
The oxidation state of hydrogen has increased from 0 to +1.

So we can say:

- **Reduction is a decrease in the oxidation state.**
- **Oxidation is an increase in the oxidation state.**
- **If oxidation states change during a reaction, it is a redox reaction.**

Variable oxidation states

- Transition elements exist in different oxidation states, in their compounds.
- So we add the oxidation state to the compound's name.
- The different oxidation states usually lead to compounds of different colours.
- So their redox reactions may give a colour change.

Transition element	Oxidation state in compounds	Examples	
copper	+I +II	Cu_2O CuO	copper(I) oxide copper(II) oxide
iron	+II +III	$FeCl_2$ $FeCl_3$	iron(II) chloride iron(III) chloride
manganese	+II +IV +VII	MnO MnO_2 Mn_2O_7	manganese(II) oxide manganese(IV) oxide manganese(VII) oxide

✓

Quick check for 11.3   *(Answers on page 167)*

1   Give the oxidation state of:
     a bromine in Br_2 **b** bromine in NaBr   **c** bromine in HBr
2    $2Mg\ (s) + CO_2\ (g) \longrightarrow 2MgO\ (s) + C\ (s)$
     a Show the oxidation state for each element in the equation.
     b Use oxidation states to show that this is a redox reaction.
     c Using oxidation states, say what is oxidised and what is reduced.
3   Using the rules above, see if you can work out the oxidation state of:
     a sulfur in sulfuric acid, H_2SO_4
     b manganese in MnO_2
     c manganese in $KMnO_4$

11.4 Oxidising and reducing agents

Look again at this redox reaction:
$$CuO\ (s) + H_2\ (g) \longrightarrow Cu\ (s) + H_2O\ (l)$$
The copper(II) oxide is being reduced by hydrogen. So hydrogen is the **reducing agent**.
The hydrogen is oxidised to water, so copper(II) oxide is the **oxidising agent**.

A reducing agent brings about reduction, and is itself oxidised.
An oxidising agent brings about oxidation, and is itself reduced.

Oxidising agents are also called **oxidants**. Reducing agents are called **reductants**.

Extended

Powerful oxidising and reducing agents

- Some substances are powerful oxidising agents, because they have a strong drive to gain electrons. Examples are oxygen, and chlorine.
- Some substances are powerful reducing agents, because they have a strong drive to give up electrons. Examples are hydrogen, carbon, carbon monoxide, and reactive metals such as sodium.

Tests for oxidising and reducing agents

When some substances are oxidised, or reduced, there is a colour change. So these substances are useful for lab tests.

Example 1: potassium manganate(VII)

Acidified potassium manganate(VII) is a powerful oxidising agent. At the same time its manganate ion is reduced, with a colour change:

$$MnO_4^- \longrightarrow Mn^{2+}$$

oxidation state	+VII	+II
colour	purple	colourless

Test: Add acidified potassium manganate(VII) solution to an unknown liquid. If a reducing agent is present, the purple colour fades.

Example 2: potassium iodide

Potassium iodide is a reducing agent. At the same time the iodide ion is oxidised, with a colour change:

$$2I^- \longrightarrow I_2$$

oxidation state	−I	0
colour	colourless	red-brown

Test: Add potassium iodide solution to an unknown liquid. If an oxidising agent is present, a red-brown colour appears.

✓

Quick check for 11.4   (Answers on page 167)

1   $2Mg (s) + CO_2 (g) \longrightarrow 2MgO (s) + C (s)$
    a Which substance is the reducing agent, in this reaction?
    b Which substance is the oxidising agent?

2   $2Fe (s) + 3Cl_2 (g) \longrightarrow 2FeCl_3 (s)$
    a Show the oxidation states, in each substance in the equation.
    b During the reaction, which substance is: i oxidised?   ii the oxidant?
    c Which substance is: i reduced?   ii the reductant?

3   Both sodium and magnesium are powerful reducing agents. Why is this?

4   What is this used to test for?
    a potassium manganate(VII)   b potassium iodide

5   Potassium dichromate(VI) is also used in lab tests. This shows the colour change:

$$Cr_2O_7^{2-} \longrightarrow 2Cr^{3+}$$

oxidation state	+VI $\longrightarrow$ +III	
colour	orange	green

    a In this change, is chromium being oxidised, or reduced?
    b So which would you test for, using potassium dichromate(VI)?
        i the presence of oxidising agents   ii the presence of reducing agents

Questions on section 11

Answers for these questions are on page 167.

Core curriculum

1 If a substance gains oxygen during a reaction, it is being oxidised. If it loses oxygen, it is being reduced. Oxidation and reduction always take place together, so if one substance is oxidised, another is reduced.

   a First, write a word equation for each redox reaction A to C below.

   b Then, using just the ideas above, say which substance is being oxidised, and which is being reduced, in each reaction.

   A $Ca\ (s) + O_2\ (g) \longrightarrow 2CaO\ (s)$

   B $2CO\ (g) + O_2\ (g) \longrightarrow 2CO_2\ (g)$

   C $Fe_2O_3\ (s) + 3CO\ (g) \longrightarrow 2Fe\ (s) + 3CO_2\ (g)$

Extended curriculum

1 Iodine is extracted from seaweed in a redox reaction using acidified hydrogen peroxide. The ionic equation for the reaction is:

$$2I^-\ (aq) + H_2O_2\ (aq) + 2H^+\ (aq) \longrightarrow I_2\ (aq) + 2H_2O\ (l)$$

   a In which oxidation state is the iodine in seaweed?

   b There is a colour change in this reaction. Why?

   c **i** Is the iodide ion oxidised or reduced, in this reaction?

        ii Complete the half equation for this change: $2I^- \longrightarrow$

   d In hydrogen peroxide, H_2O_2, the oxidation state of the hydrogen is +I.

        i What is the oxidation state of the oxygen in hydrogen peroxide?

        ii How does the oxidation state of oxygen change during the reaction above?

        iii Complete the half equation for hydrogen peroxide:

          $H_2O_2\ (aq) + 2H^+\ (aq) + \ldots\ldots \longrightarrow 2H_2O\ (l)$

2 The oxidising agent potassium manganate(VII) can be used to analyse the % of iron(II) present in iron tablets. This ionic equation shows the ions taking part in the reaction:

$$MnO_4^-\ (aq) + 8H^+\ (aq) + 5Fe^{2+}\ (aq) \longrightarrow Mn^{2+}\ (aq) + 5Fe^{3+}\ (aq) + 4H_2O\ (l)$$

   a What does the H^+ in the equation tell you about this reaction?

   b Describe the colour change.

   c Which is the reducing reagent in this reaction?

   d How could you tell when all the iron(II) had reacted?

   e Write a half equation to show how the iron(II) reacts.

3 Cells can be set up with inert electrodes, and the electrolytes as oxidant and reductant, as shown on the right.

Here potassium manganate(VII) is the oxidant and potassium iodide is the reductant.

   a Describe the colour change that would be observed in the left hand beaker.

   b Write an ionic equation for the reaction in the right hand beaker.

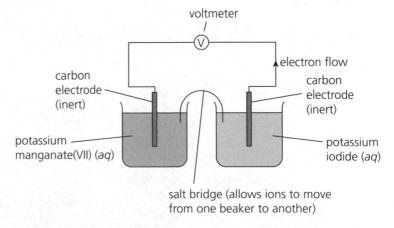

CIE 0620 June '06 Paper 3 Q6

12 Acids and bases

The big picture

- Acids and bases are two very important groups of compounds.
- Acids react with metals, bases, and carbonates, to form a wide range of compounds called salts.
- In many of the reactions of acids, a salt and water are formed. Those reactions are called neutralisations.

12.1 Acids and alkalis

Some acids

hydrochloric acid	HCl (*aq*)
sulfuric acid	H_2SO_4 (*aq*)
nitric acid	HNO_3 (*aq*)

Alkalis (or soluble bases)

sodium hydroxide	NaOH (*aq*)
potassium hydroxide	KOH (*aq*)
ammonia solution	NH_3 (*aq*)

The pH scale

The pH scale is used to show how acidic or alkaline a solution is. It goes from 0 to 14.

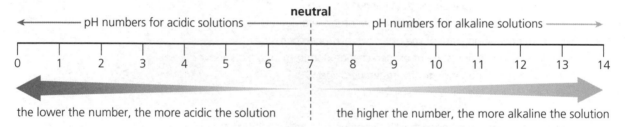

the lower the number, the more acidic the solution   the higher the number, the more alkaline the solution

We use universal indicator, or a pH meter, to find the pH of a solution.
This shows the colour of universal indicator at different pH values:

pH	1	2	3	4	5	6	7	8	9	10	11	12	13	14
colour	red		orange		yellow		green			blue		purple-violet		

Other indicators

Indicator	Colour in acid	Colour in alkali
litmus	red	blue
phenolpthalein	colourless	pink
methyl orange	red	yellow

Quick check for 12.1 *(Answers on page 167)*

1 Here are the pH values for some solutions.

   A pH 6   **B** pH 10   **C** pH 1   **D** pH 7   **E** pH 13

   a For each solution, say whether it is acidic, alkaline, or neutral.

   b Then pick out the most acidic and most alkaline solutions.

2 What colour does universal indicator show, in a neutral solution?

12.2 The reactions of acids

Acids react with
- metals
- bases (metal oxides, hydroxides, and ammonia)
- carbonates

to form **salts**.

<table>
<tr><td colspan="2">Naming salts</td></tr>
<tr><td>acid</td><td>salts</td></tr>
<tr><td>nitric</td><td>nitrates</td></tr>
<tr><td>sulfuric</td><td>sulfates</td></tr>
<tr><td>hydrochloric</td><td>chlorides</td></tr>
</table>

Examples of acid reactions

Acid + ...	Product	Example
metal	salt + hydrogen	magnesium + hydrochloric acid $\longrightarrow$ magnesium chloride + hydrogen $Mg\ (s)\ +\ 2HCl\ (aq)\ \longrightarrow\ MgCl_2\ (aq)\ +\ H_2\ (g)$
hydroxide	salt + water	sodium hydroxide + nitric acid $\longrightarrow$ sodium nitrate + water $NaOH\ (s)\ +\ HNO_3\ (aq)\ \longrightarrow\ NaNO_3\ (aq)\ +\ H_2O\ (l)$
metal oxide	salt + water	copper(II) oxide + sulfuric acid $\longrightarrow$ copper(II) sulfate + water $CuO\ (s)\ +\ H_2SO_4\ (aq)\ \longrightarrow\ CuSO_4\ (aq)\ +\ H_2O\ (l)$
carbonate	salt + water + carbon dioxide	calcium carbonate + nitric acid $\longrightarrow$ calcium nitrate + water + carbon dioxide $CaCO_3\ (s)\ +\ 2HNO_3\ (aq)\ \longrightarrow\ Ca(NO_3)_2\ (aq)\ +\ H_2O\ (l)\ +\ CO_2\ (g)$

In the last three reactions above, a **salt** and **water** are formed. So they are **neutralisations**.

✓ **Quick check for 12.2**   (Answers on page 167)
1. In a reaction with acid, hydrogen is produced.
   a What has the acid reacted with?
   b Write a chemical equation for an example of this reaction.
2. Write the chemical equation for the reaction between magnesium oxide and hydrochloric acid.
3. Hydrochloric acid is added to calcium carbonate.
   a Name the gas given off in the reaction.
   b Write the balanced symbol equation for the reaction.

12.3 Bases and their reactions

- The bases are metal oxides and hydroxides, and ammonia.
- The **soluble** bases are called **alkalis**.
- Bases neutralise acids, forming a salt and water.
- Bases (other than ammonia) displace ammonia from ammonium salts.

Examples of base reactions

Base + ...	Product	Example
acid (*neutralisation*)	salt + water	potassium hydroxide + nitric acid $\longrightarrow$ potassium nitrate + water $KOH\ (s)\ +\ HNO_3\ (aq)\ \ \ \ KNO_3\ (aq)\ +\ H_2O\ (l)$
acid (*neutralisation*)	salt + water	calcium oxide + hydrochloric acid $\longrightarrow$ calcium chloride + water $CaO\ (s)\ +\ 2HCl\ (aq)\ \ \ \ CaCl_2\ (aq)\ +\ H_2O\ (l)$
ammonium compound (*displacement*)	salt + water + ammonia	calcium hydroxide + ammonium chloride $\longrightarrow$ calcium chloride + water + ammonia $Ca(OH)_2\ (s)\ +\ 2NH_4Cl\ (aq)\ \ \ \ CaCl_2\ (aq)\ +\ 2H_2O\ (l)\ +\ 2NH_3\ (g)$

Using bases to control acidity in soil

- Soil can be acidic or alkaline. Its pH depends on factors such as the rock it came from, the decayed vegetation in it, the fertilisers used on it, and whether there is acid rain.
- Crops and other plants take in nutrients such as nitrogen, phosphorus, and potassium from the soil, through their roots.
- But if the soil is too acidic, the nutrients are not available in a form the plants can use. (The best pH for most crops is neutral to slightly acidic.)
- So calcium oxide, calcium hydroxide, and calcium carbonate are spread on soil that is too acid, to neutralise its acidity.
- These compounds are chosen because they are quite cheap, and only sparingly (slightly) soluble in water, so rain won't wash them away.

Make the link to ...
limestone, quicklime, and slaked lime, on page 136.

✓

Quick check for 12.3 *(Answers on page 167)*

1 How would you obtain ammonia from an ammonium salt?
2 Explain why:
   a crops will not grow well if soil is too acidic
   b farmers spread crushed limestone (calcium carbonate) on their soil.

12.4 A closer look at acids and bases

Extended

H$^+$ and OH$^-$ ions

- All solutions of acids contain hydrogen ions. It is these that make them 'acidic'.
  For example hydrochloric acid **ionises** (forms ions) in water, like this:
  $HCl\,(aq) \longrightarrow H^+\,(aq) + Cl^-\,(aq)$
- All solutions of alkalis contain hydroxide ions. It is these that make them alkaline.
  For example the ions in sodium hydroxide separate like this:
  $NaOH\,(aq) \longrightarrow Na^+\,(aq) + OH^-\,(aq)$

Strong and weak acids and bases

Strength is about the extent to which H$^+$ and OH$^-$ ions are formed.

Acids		Alkalis (soluble bases)	
Strong acids exist completely as ions, in water. For example in hydrochloric acid: $HCl\,(aq) \xrightarrow{100\%} H^+\,(aq) + Cl^-\,(aq)$		**Strong alkalis** exist completely as ions, in water. For example in sodium hydroxide: $NaOH\,(aq) \xrightarrow{100\%} Na^+\,(aq) + OH^-\,(aq)$	
Weak acids exist only partly as ions, in water. For example in ethanoic acid: $CH_3COOH\,(aq) \xrightarrow{much\ less\ than\ 100\%} H^+\,(aq) + CH_3COO^-\,(aq)$		Ammonia solution is a **weak alkali**, because only some ammonia molecules form ions: $NH_3\,(aq) \xrightarrow{much\ less\ than\ 100\%} NH_4^+\,(aq) + OH^-\,(aq)$	
A 0.1 M solution of hydrochloric acid contains many more H$^+$ ions than a 0.1M solution of ethanoic acid. So it is more acidic. It has a lower pH.		A 0.1 M solution of sodium hydroxide contains many more OH$^-$ ions than a 0.1M solution of ammonia. So it is more alkaline. It has a higher pH.	
Strong acids hydrochloric acid sulfuric acid nitric acid	**Weak acids** ethanoic acid citric acid	**Strong alkalis** sodium hydroxide potassium hydroxide	**Weak alkali** ammonia solution

But note that it takes the same amount of alkali to neutralise a 0.1M solution of ethanoic acid as a 0.1 M solution of hydrochloric acid. That is because, as its H$^+$ ions get 'used up' in the neutralisation, the ethanoic acid continues to ionise.

The neutralisation reaction

Look at the neutralisation of hydrochloric acid by sodium hydroxide solution:

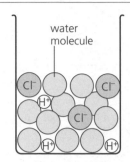	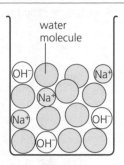	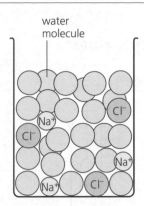
A solution of hydrochloric acid contains H^+ ions and Cl^- ions. It will turn litmus red.	A solution of sodium hydroxide contains Na^+ ions and OH^- ions. It will turn litmus blue.	When you mix the two solutions, the H^+ and OH^- ions join to form water molecules. The result is a neutral solution containing Na^+ and Cl^- ions. It has no effect on litmus.

So the Na^+ and Cl^- ions remain unchanged. Only the H^+ and OH^- ions react.

In a neutralisation reaction, H^+ and OH^- ions combine to form water molecules:

$$H^+ + OH^- \longrightarrow H_2O$$

Spectator ions

The Na^+ ions and Cl^- ions are present at the neutralisation reaction above, but they do not take part in it. So they are called **spectator ions**.

Proton donors and acceptors

Note that the H^+ ion is just a proton, as this drawing shows:

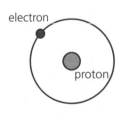

electron

proton

a hydrogen atom

 proton

a hydrogen ion is just a proton

So in the neutralisation reaction above, the hydrochloric acid has *donated* protons.
The hydroxide ions from sodium hydroxide have *accepted* them.
So this gives us a new definition for acids and bases:
Acids are proton donors. Bases are proton acceptors.

> **Remember**
> Anything that accepts a proton in a reaction is acting as a base.

> **Remember**
> Anything that donates a proton in a reaction is acting as an acid.

✓
Quick check for 12.4   *(Answers on page 167)*
1   A solution contains hydroxide ions. What can you say about its pH?
2   What is the key difference between a strong acid and a weak acid?
3   Name: **a** a strong acid   **b** a weak acid   **c** a weak alkali
4   Which of these has a higher pH?
     a a 1M solution of sodium hydroxide   **b** a 1M solution of ammonia
5   What is another name for a hydrogen ion?
6   What does the equation $H^+ (aq) + OH^- (aq) \longrightarrow H_2O (l)$ represent?
7   Complete this sentence: *acids are proton* ....

12.5 Oxides

Oxygen is very reactive. Its compounds with other elements are called **oxides**.

Basic oxides

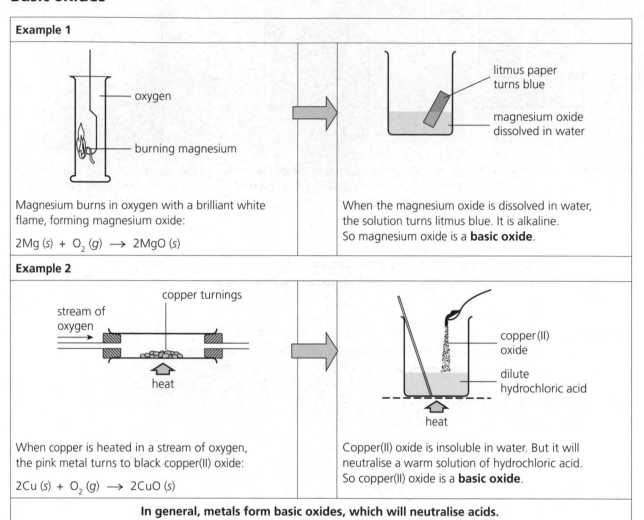

Example 1

Magnesium burns in oxygen with a brilliant white flame, forming magnesium oxide:

$2Mg\ (s)\ +\ O_2\ (g)\ \longrightarrow\ 2MgO\ (s)$

When the magnesium oxide is dissolved in water, the solution turns litmus blue. It is alkaline. So magnesium oxide is a **basic oxide**.

litmus paper turns blue

magnesium oxide dissolved in water

Example 2

oxygen

burning magnesium

stream of oxygen

copper turnings

heat

When copper is heated in a stream of oxygen, the pink metal turns to black copper(II) oxide:

$2Cu\ (s)\ +\ O_2\ (g)\ \longrightarrow\ 2CuO\ (s)$

copper(II) oxide

dilute hydrochloric acid

heat

Copper(II) oxide is insoluble in water. But it will neutralise a warm solution of hydrochloric acid. So copper(II) oxide is a **basic oxide**.

In general, metals form basic oxides, which will neutralise acids.

For example, the reaction between copper(II) oxide and hydrochloric acid is:

copper(II) oxide + hydrochloric acid $\longrightarrow$ copper(II) chloride + water

$CuO\ (s)\quad +\qquad 2HCl\ (aq)\qquad \longrightarrow\qquad CuCl_2\ (aq)\quad +\ H_2O\ (l)$

> **Some other basic oxides:**
> sodium oxide, Na_2O
> calcium oxide, CaO
> iron(III) oxide, Fe_2O_3

Acidic oxides

The oxides of non-metals behave in a different way from the metal oxides above.

Example Sulfur catches fire over a Bunsen. It burns with a blue flame, forming sulfur dioxide:

$S\ (s)\ +\ O_2\ (g)\ \longrightarrow\ SO_2\ (g)$

This gas is soluble in water. The solution turns litmus red. So sulfur dioxide is an **acidic oxide.**

In general, non-metals form acidic oxides. These will neutralise bases.

For example, the reaction between sulfur dioxide and sodium hydroxide is:

sulfur dioxide + sodium hydroxide $\longrightarrow$ sodium sulfite + water

$SO_2\ (g)\qquad +\qquad 2NaOH\ (aq)\qquad \longrightarrow\qquad Na_2SO_3\ (aq)\ +\ H_2O\ (l)$

> **Some other acidic oxides:**
> carbon dioxide, CO_2
> nitrogen dioxide, NO_2
> phosphorus pentoxide, P_4O_{10}

hoteric oxides

oxides are both acidic *and* basic. They are called **amphoteric**.

ple Aluminium burns in oxygen with a white flame, to form aluminium oxide:

$+ 3O_2 (g) \longrightarrow Al_2O_3 (s)$

ium oxide neutralises acid, to form a salt. So it acts as a basic oxide. For example:

$(s) + 6HCl (aq) \longrightarrow 2AlCl_3 (aq) + 3H_2O (l)$
aluminium chloride

lso reacts with dilute alkali, to form a salt. So it acts as an acidic oxide. For example:

$(s) + 6NaOH (aq) \longrightarrow 2Na_3AlO_3 (aq) + 3H_2O (l)$
sodium aluminate

minium oxide is an amphoteric oxide.

Some other amphoteric oxides:
zinc oxide, ZnO
lead(II) oxide, PbO

tral oxides

xides are neither acidic *nor* basic. They are called **neutral**.

le Carbon burns in a limited supply of oxygen to form carbon monoxide:

$+ O_2 (g) \longrightarrow 2CO (g)$

monoxide:

s no effect on litmus
es not react with dilute acid, so it is not basic
es not react with dilute alkali, so it is not acidic.

on monoxide is a neutral oxide.

Remember

The pH of water and neutral solutions is 7.

Some other neutral oxides:
water, H_2O
dinitrogen oxide, N_2O

---

ick check for 12.5   (*Answers on page 167*)

Which type of element usually forms acidic oxides?
Suggest a compound to react with copper(II) oxide, to show that it is a basic oxide.
Will sodium hydroxide react with: **a** magnesium oxide? **b** carbon dioxide?
Explain your answer each time.
When sulfur dioxide and nitrogen dioxide escape into the air from power station chimneys, the result is acid rain. Explain why.
What evidence would you give, to show that zinc oxide is amphoteric?
Support your answer with suitable chemical equations.
Dinitrogen oxide is a neutral oxide. What does that mean?

Questions on section 12

Answers for these questions are on page 167.

Core curriculum

1 A and B are white powders. A is insoluble in water. B forms a solution with a pH of 3.
   A mixture of A and B effervesces in water, giving off a gas. A clear solution forms.
   a One of the white powders is an acid. Which one?
   b The other powder is a carbonate. Which gas bubbles off in the reaction?
   c Although A is insoluble in water, a clear solution forms when the mixture of A and
      B is added to water. Explain why.

2 Pure dry crystals of magnesium sulfate can be made by reacting excess magnesium
   powder with dilute sulfuric acid.
   a During the reaction, bubbles of a colourless gas are given off.
      State the name of this gas.
   b i Why is excess magnesium used?
      ii How is the excess magnesium removed from the reaction mixture?
   c Describe how you can obtain pure dry crystals of magnesium sulfate from a
      solution of magnesium sulfate.
   d i Describe one other reaction that makes magnesium sulfate.
      ii Write a word equation for the reaction you suggested in part **d i**.
      iii Magnesium sulfate can be used as a medicine. Explain why the chemicals used
          in medicines need to be as pure as possible.0

CIE 0620 June '08 Paper 2 Q5

3 The diagram shows the changes in pH in a student's mouth after she has eaten a sweet.

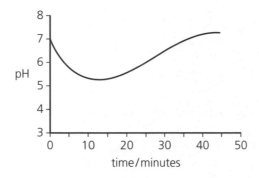

   a Describe how the acidity in the student's mouth changes after eating the sweet.
   b i Chewing a sweet stimulates the formation of saliva. Saliva is slightly alkaline.
         Use this information to explain the shape of the graph.
      ii Name the type of reaction which occurs when an acid reacts with an alkali.
   c Many sweets contain citric acid.
      Citric acid can be extracted from lemon juice as follows:
      stage 1: add calcium carbonate to hot lemon juice
      stage 2: filter off the precipitate which is formed (calcium citrate)
      stage 3: wash the calcium citrate precipitate with water
      stage 4: add sulfuric acid to the calcium citrate to make a solution of citric acid
      stage 5: crystallise the citric acid
      i When calcium carbonate is added to lemon juice a fizzing is observed.
         Explain why there is a fizzing.
      ii Draw a diagram to show step 2. Label your diagram.
      iii Suggest why the calcium citrate precipitate is washed with water.
      iv Describe how you would carry out step 5.

CIE 0620 November '07 Paper 2 Q4

4   Seven oxides are listed below.

calcium oxide   carbon dioxide   carbon monoxide
phosphorus trioxide   sodium oxide   copper(II) oxide
water

a   Which one of these oxides is an insoluble base?
b   Which one of these oxides is a product of the reaction between an acid and
     a carbonate?
c   Which one of these oxides is formed when carbon is burnt in a limited supply
     of air?
d   Which one of these oxides is a good solvent?
e   Which one of these oxides is a neutral oxide?
f   Which **two** of these oxides react with water to form an alkaline solution?
g   Which **two** of these oxides form acidic solutions when they dissolve in water?

CIE 0620 November '07 Paper 2 Q1

Extended curriculum

1   Sulfuric acid is a typical **strong** acid.
a   Change the equations given into a different format.
   i $Mg + H_2SO_4 \longrightarrow MgSO_4 + H_2$
        Change this into a word equation.
   ii   lithium oxide + sulfuric acid $\longrightarrow$ lithium sulfate + water
        Change this into a symbol equation.
   iii $CuO + 2H^+ \longrightarrow Cu^{2+} + H_2O$
        Change the ionic equation into a symbol equation.
   iv $Na_2CO_3 + H_2SO_4 \longrightarrow Na_2SO_4 + CO_2 + H_2O$
        Change this into a word equation.
b   When sulfuric acid dissolves in water, the following reaction occurs.
     $H_2SO_4 + H_2O \longrightarrow HSO_4^- + H_3O^+$
     Explain why water is behaving as a base in this reaction.
c   Sulfuric acid is a strong acid, ethanoic acid is a weak acid.
     Explain the difference between a strong acid and a weak acid.

CIE 0620 June '08 Paper 3 Q4

2   Methylamine, CH_3NH_2, is a weak base. Its properties are similar to those of ammonia.
a   When methylamine is dissolved in water, the following equilibrium is set up.
     $CH_3NH_2 + H_2O \rightleftharpoons CH_3NH_3^+ + OH^-$
       base   acid
   i   Suggest why the arrows are not the same length.
   ii   Explain why water is stated to behave as an acid and methylamine as a base.
b   An aqueous solution of the strong base, sodium hydroxide, is pH 12. Predict the pH
     of an aqueous solution of methylamine which has the same concentration. Give a
     reason for your choice of pH.
c   Methylamine is a weak base like ammonia.
   i   Methylamine can neutralise acids.
        $2CH_3NH_2$ $+$ H_2SO_4 $\longrightarrow$ $(CH_3NH_3)_2SO_4$
                                                                            methylammonium sulfate
        Write the equation for the reaction between methylamine and hydrochloric
        acid. Name the salt formed.
   ii   When aqueous methylamine is added to aqueous iron(II) sulfate, a green
        precipitate is formed. What would you see if iron(III) chloride solution had been
        used instead of iron(II) sulfate?
   iii Suggest a reagent to displace methylamine from one of its salts, for example
        methylammonium sulfate.

CIE 0620 November '07 Paper 3 Q5

13 Preparing salts

The big picture

- Salts are a very important group of compounds, with many uses.
- To make a salt in the lab, the first task is to choose suitable reactants.
- The method used to make the salt depends on the solubility of the reactants and the salt.

> **Note**
> 'Reactant' and 'reagent' mean the same thing – a substance that is reacting.

13.1 Making a soluble salt in the lab

Which salts are soluble?

The method you use to make a salt depends on whether the salt is soluble.

This is a useful list of soluble salts:

- all sodium, potassium and ammonium salts
- all nitrates
- all chlorides *except silver and lead chlorides*
- all sulfates *except calcium, barium and lead sulfates*
- sodium, potassium and ammonium carbonates *but all other carbonates are insoluble.*

Making a soluble salt from an acid

The list of soluble salts above includes all nitrates, and most chlorides and sulfates.

You make these from the corresponding acids. This flowchart summarises the process:

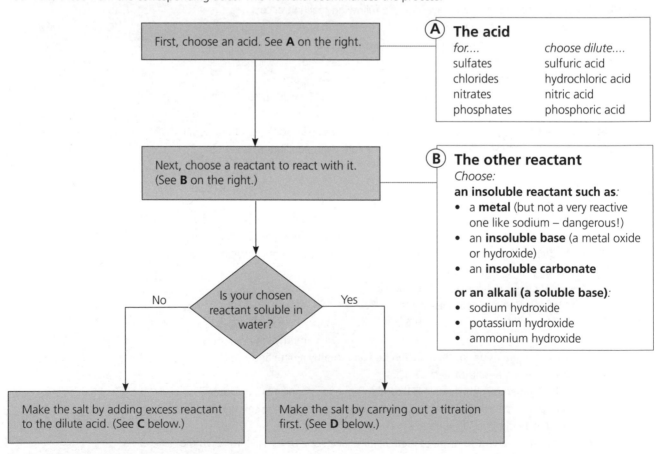

First, choose an acid. See **A** on the right.

(A) The acid

for....	choose dilute....
sulfates	sulfuric acid
chlorides	hydrochloric acid
nitrates	nitric acid
phosphates	phosphoric acid

Next, choose a reactant to react with it. (See **B** on the right.)

(B) The other reactant

Choose:

an insoluble reactant such as:
- a **metal** (but not a very reactive one like sodium – dangerous!)
- an **insoluble base** (a metal oxide or hydroxide)
- an **insoluble carbonate**

or an alkali (a soluble base):
- sodium hydroxide
- potassium hydroxide
- ammonium hydroxide

Is your chosen reactant soluble in water?

No → Make the salt by adding excess reactant to the dilute acid. (See **C** below.)

Yes → Make the salt by carrying out a titration first. (See **D** below.)

C The preparation using an insoluble reactant

If you choose an insoluble reactant, follow these steps:

1 Allow the reactants to react.	2 Filter the solution.
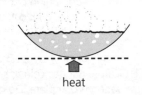 — dilute acid — solid reagent	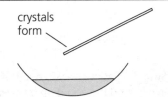 — unreacted solid — aqueous solution of the salt
Use an **excess** of the solid to make sure all the acid reacts.	Unreacted solid remains in the filter paper. The filtrate passes through.

3 Obtain salt crystals from the solution.

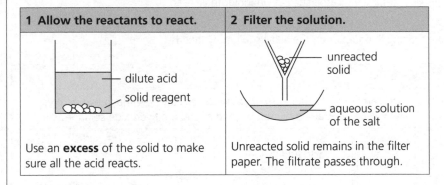 heat	crystals form	crystals start to form
Heat the filtrate … to evaporate some of the water, in order to obtain a **saturated solution**.	**Check …** using a glass rod. If crystals form on the cool rod, the solution is saturated. You can stop heating.	**Leave to crystallise.** As the solution cools, pure crystals of the salt form. Filter them off, and dry them.

Why the crystals form

The crystals form, in step 3 above, because water is able to dissolve less and less solute as its temperature falls. The check with the glass rod tells you whether the solution is saturated. (A solution is saturated if the solvent can dissolve no more solute, at that temperature.) As the saturated solution cools down, crystals form.

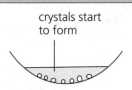

Remember

Some salts are broken down by heat – so should not be dried in an oven. Use filter paper instead.

Examples of the reactions, using an insoluble reactant

1 Metal + acid

zinc + sulfuric acid ⟶ zinc sulfate + hydrogen

$Zn\ (s) + H_2SO_4\ (aq) \longrightarrow ZnSO_4\ (aq) + H_2\ (g)$

You see fizzing as hydrogen gas is given off. You know the reaction is complete when there is still some zinc left, but no more fizzing (because all the acid has been used up).

2 Insoluble base + acid

You can heat the mixture to speed up the reaction. For example:

copper(II) oxide + sulfuric acid $\xrightarrow{\text{heat}}$ copper(II) sulfate + water

$CuO\ (s) + H_2SO_4\ (aq) \longrightarrow CuSO_4\ (aq) + H_2O\ (l)$

You see no fizzing, but the solution turns blue as the solid dissolves. You know the reaction is complete when no more solid will dissolve (because all the acid has been used up), and there is no further colour change.

3 Insoluble carbonate + acid

magnesium carbonate + nitric acid ⟶ magnesium nitrate + water + carbon dioxide

$MgCO_3\ (s) + 2HNO_3\ (aq) \longrightarrow Mg(NO_3)_2\ (aq) + H_2O\ (l) + CO_2\ (g)$

You see fizzing as carbon dioxide gas is given off. You know the reaction is complete when there is still some magnesium carbonate left, but no more fizzing (because all the acid has been used up).

D The preparation using a soluble reactant (an alkali)

You must do a **titration** first, to find the exact volume of acid that will be neutralised by a known volume of alkali. Use a suitable acid-base indicator, such as phenolphthalein, to show when neutralisation is complete.

For example this shows how to make sodium chloride, using the alkali sodium hydroxide:

1 Titration, to see how much acid is needed.		
Add indicator to a known volume of sodium hydroxide solution.	Then carry out a titration as below, to see exactly how much hydrochloric acid will be neutralised by that volume of sodium hydroxide solution.	
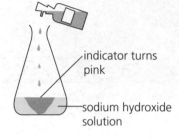 indicator turns pink — sodium hydroxide solution	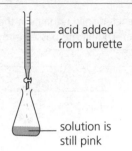 acid added from burette — solution is still pink	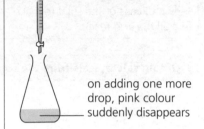 on adding one more drop, pink colour suddenly disappears
	Swirl the flask while you drip the acid into it. Add the acid drop by drop when you think the end-point of the neutralisation is near.	The indicator changes colour, so neutralisation is complete. Do not add any more acid. Note down the volume of acid you used.

2 Make the salt	3 Obtain the salt crystals
Repeat the titration **without** the indicator. Add the same volume of acid as you did before.	Obtain sodium chloride crystals by heating the solution carefully, to evaporate all the water.
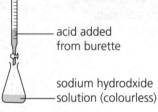 acid added from burette — sodium hydrodxide solution (colourless)	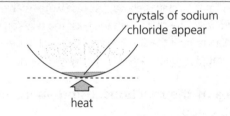 crystals of sodium chloride appear — heat
There is no need for indicator now, since you know how much acid to add. The indicator would be an unwanted impurity.	You must evaporate all the water, since the solubility of sodium chloride does not change much as temperature falls. (But for other salts, see step 3 on page 93.)

The equation for the neutralisation above is:

sodium hydroxide + hydrochloric acid ⟶ sodium chloride + water

$NaOH\ (aq)$ + $HCl\ (aq)$ ⟶ $NaCl\ (aq)$ + $H_2O\ (l)$

> ✓
> **Quick check for 13.1** *(Answers on page 167)*
> 1. Suggest reactants for making these soluble salts:
>   a magnesium nitrate   b zinc sulfate   c zinc chloride
> 2. List the steps in making magnesium chloride, from magnesium and an acid.
> 3. a It would be a bad idea to make calcium chloride by starting with calcium metal. Why?
>   b Suggest suitable reactants for making calcium chloride.
> 4. What exactly is the purpose of an indicator, in a titration?
> 5. Name three different salts you could make, by titration.

13.2 Making an insoluble salt in the lab

- A **precipitate** is an insoluble chemical produced during a chemical reaction.
- You make an insoluble salt in the lab by **precipitation**.
- You mix a solution containing its positive ions with one containing its negative ions.

> **Insoluble salts**
> **chlorides**: silver and lead
> **sulfates**: calcium, barium, lead
> **carbonates**: all except sodium, potassium and ammonium

Example: making silver chloride

The insoluble salt silver chloride contains silver ions (Ag^+) and chloride ions (Cl^-).

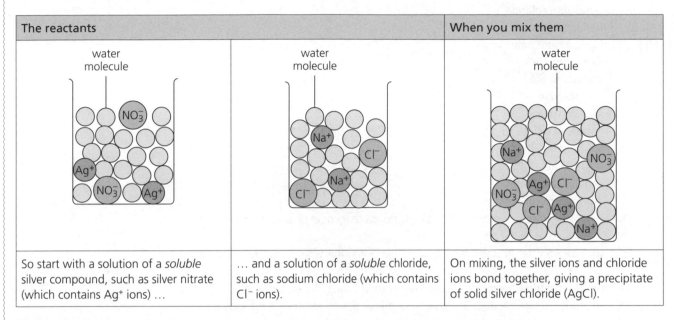

The reactants		When you mix them
So start with a solution of a *soluble* silver compound, such as silver nitrate (which contains Ag^+ ions) …	… and a solution of a *soluble* chloride, such as sodium chloride (which contains Cl^- ions).	On mixing, the silver ions and chloride ions bond together, giving a precipitate of solid silver chloride (AgCl).

The equation for the reaction above is:

silver nitrate + sodium chloride $\longrightarrow$ silver chloride + sodium nitrate

$AgNO_3$ (aq) + NaCl (aq) $\longrightarrow$ AgCl (s) + $NaNO_3$ (aq)

Look at the state symbols in the equation. Only the silver chloride is solid.

Spectator ions

- In the reaction above, only the silver ions and chloride ions react together.
- The nitrate ions and the sodium ions remain in solution unchanged – so they are called **spectator ions**. (They just sit and watch!)
- An **ionic equation** does not include the spectator ions. It shows what actually happens in the precipitation reaction. The ionic equation for this reaction is:
  Ag^+ (aq) + Cl^- (aq) $\longrightarrow$ AgCl (s)

The method for making salts by precipitation

1. Add an excess of one solution to the other. A precipitate forms. It is the salt.
2. Filter the liquid to separate the precipitate. You can discard the filtrate.
3. Wash the precipitate with distilled water to remove unwanted reactant.
4. Dry off in a warm oven. (The oven must not be hot enough to decompose the salt!)

solution of
second reactant

solution of
reactant

precipitate
forms

Making a precipitate: step 1

✓
Quick check for 13.2   (*Answers on page 167*)
1. What does *precipitation* mean?
2. Write: **a** a word equation   **b** a symbol equation   **c** an ionic equation
   for the reaction between silver nitrate solution and calcium chloride solution.
3. Name the spectator ions for the reaction in question **2**.
4. Suggest reactants for making these insoluble salts, by precipitation:
   a lead chloride   **b** calcium sulfate   **c** magnesium carbonate

Questions on section 13

Answers for these questions start on page 167.

Core curriculum

1 **a** Write two lists, dividing these salts into *soluble in water*, and *insoluble in water*.

sodium chloride

calcium carbonate

potassium chloride

barium sulfate

barium carbonate

silver chloride

sodium citrate

zinc chloride

sodium sulfate

copper(II) sulfate

lead sulfate

lead nitrate

sodium carbonate

ammonium carbonate

   b **i** Now write down two starting compounds that could be used to make each insoluble salt.

   ii For each set of starting compounds, say what this type of reaction is called.

2 25 cm³ of potassium hydroxide solution were put in a flask. A few drops of phenolphthalein were added. Dilute hydrochloric acid was added until the indicator changed colour. It was found that 21 cm³ of acid was used.

   a Draw a labelled diagram of apparatus suitable for this titration.

   b What apparatus is used to measure 25.0 cm³ of solution accurately?

   c What is the phenolphthalein for?

   d What colour was the solution in the flask at the start of the titration?

   e What colour did it turn when the alkali had been neutralised?

   f Which solution was more concentrated, the acid or the alkali? Explain your answer.

   g Name the salt formed in this neutralisation.

   h Write a balanced equation for the reaction.

   i How would you obtain pure crystals of the salt?

Extended curriculum

1 There are three methods of preparing salts.

Method A – use a burette and an indicator.

Method B – mix two solutions and obtain the salt by precipitation.

Method C – add an excess of base or a metal to a dilute acid and remove the excess by filtration.

For each of the salt preparations **i** to **iii** below:

   a choose one of the methods A, B or C,

   b name any additional reagent needed,

   c write the equation for the chemical reaction.

   i the **soluble** salt, zinc sulfate, from the insoluble base, zinc oxide

   ii the **soluble** salt, potassium chloride, from the soluble base, potassium hydroxide

   iii the **insoluble** salt, lead(II) iodide, from the soluble salt, lead(II) nitrate

CIE 0620 June '07 Paper 3 Q3

2   There are three ways of making salts from sulfuric acid:
    titration using a burette and indicator
    precipitation by mixing the solutions and filtering
    neutralisation of sulfuric acid using an excess of an insoluble base.
    Complete the following table of salt preparations.

Method	Reactant 1	Reactant 2	Salt
titration	sulfuric acid	a	sodium sulfate
neutralisation	sulfuric acid	b	zinc sulfate
precipitation	sulfuric acid	c	barium sulfate
d	sulfuric acid	copper(II) oxide	copper(II) sulfate

CIE 0620 November '02 Paper 3 Q1

Alternative to practical

1   A solution of copper sulfate was made by reacting excess copper oxide with dilute
    sulfuric acid. The diagram shows the method used.

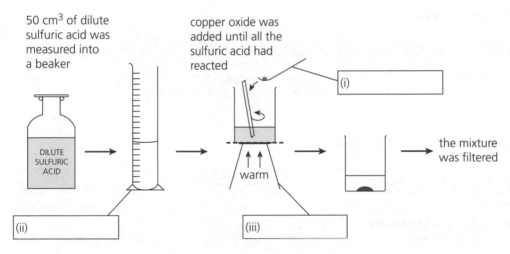

50 cm³ of dilute sulfuric acid was measured into a beaker

copper oxide was added until all the sulfuric acid had reacted

(i)

DILUTE SULFURIC ACID

warm

the mixture was filtered

(ii)   (iii)

a   Name the three pieces of apparatus (i – iii).
b   What does the term *excess* mean?
c   Draw a labelled diagram to show how the mixture was filtered.

CIE 0620 June '08 Paper 6 Q1

2   The information in the box is about the preparation of zinc nitrate crystals.

Step 1: Add a small amount of zinc oxide to some hot dilute nitric acid, and stir.
Step 2: Keep adding zinc oxide until it is in excess.
Step 3: Remove the excess zinc oxide to leave colourless zinc nitrate solution.
Step 4: Evaporate the zinc nitrate solution until it is saturated.
Step 5: Leave the saturated solution to cool. White crystals form on cooling.
Step 6: Remove the crystals from the remaining solution.
Step 7: Dry the crystals on a piece of filter paper.

a   Suggest a reason for using excess zinc oxide in Step 2.
b   Suggest how the excess zinc oxide can be removed from the solution in Step 3.
c   i   What is meant by the term *saturated solution*?
    ii   What practical method could show the solution to be saturated?
d   Why are the crystals dried in Step 7 using filter paper instead of by heating?

CIE 0620 November '07 Paper 6 Q3

14 Identifying ions and gases in the lab

The big picture

- In the laboratory, you may have to be a detective, and test substances to find out what they are.
- There are standard tests you can use, to test for the presence of different ions, and gases.
- The tests for ions rely on reactions that form a precipitate, or release a gas.

14.1 The principles of ion testing

Suppose you have a solution of an unknown ionic compound in water.
You have to work out what the compound is. This information will help you prepare.

First, ions fall into two groups:

Cations (positive ions)	Anions (negative ions)
• metal ions (such as Na^+, Al^{3+}) • ammonium ions (NH_4^+)	• halide ions (from Group VII, such as Cl^-) • compound ions containing oxygen and another non-metal (such as NO_3^-, SO_4^{2-})

Second, there are two methods of testing:

By forming a precipitate	By releasing a gas
1 Add a chemical that reacts with the unknown ion to form an insoluble compound or **precipitate**. 2 Observe the colour of the precipitate. 3 If necessary, add further chemicals to see how the precipitate behaves.	1 Add a chemical that reacts with the unknown ion, to release a gas. 2 Identify the gas, by observation and tests.

All the tests for ions that follow use one of these two methods. Many make use of reactions that you have met in other parts of your chemistry course.

14.2 Testing for cations

Testing for ammonium ions (NH_4^+)

By releasing a gas. Bases displace ammonia from ammonium compounds. (See 12.3.)
So add sodium hydroxide to the unknown compound. If ammonia is given off,
the NH_4^+ ion is present. The steps are:

1 Take a small amount of the unknown solid or solution. Add a small amount of dilute sodium hydroxide solution, and heat gently.
2 If ammonia gas is given off, the unknown substance contained ammonium ions.
   (Ammonia turns damp red litmus paper blue.)
   The reaction that takes place is:

$$NH_4^+ (aq) \ + \ OH^- (aq) \ \longrightarrow \ NH_3 (g) \ + \ H_2O (l)$$

from sodium hydroxide   ammonia

> **Remember**
> Cations (+) would go to the cathode (–) during electrolysis.

Testing for metal ions (Cu^{2+}, Fe^{2+}, Fe^{3+}, Al^{3+}, Zn^{2+}, Ca^{2+})

By forming a precipitate. The hydroxides of all the above metals are insoluble. So if you add sodium hydroxide to a solution containing any of their ions, a precipitate will form.

A flowchart for the tests

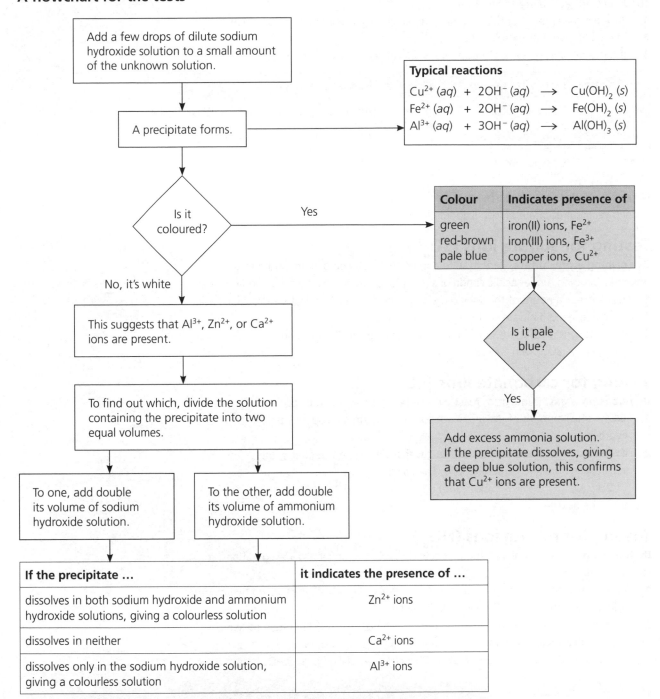

Typical reactions

$$Cu^{2+} (aq) + 2OH^- (aq) \longrightarrow Cu(OH)_2 (s)$$
$$Fe^{2+} (aq) + 2OH^- (aq) \longrightarrow Fe(OH)_2 (s)$$
$$Al^{3+} (aq) + 3OH^- (aq) \longrightarrow Al(OH)_3 (s)$$

Add a few drops of dilute sodium hydroxide solution to a small amount of the unknown solution.

A precipitate forms.

Is it coloured?

Yes

Colour	Indicates presence of
green	iron(II) ions, Fe^{2+}
red-brown	iron(III) ions, Fe^{3+}
pale blue	copper ions, Cu^{2+}

No, it's white

This suggests that Al^{3+}, Zn^{2+}, or Ca^{2+} ions are present.

Is it pale blue?

Yes

To find out which, divide the solution containing the precipitate into two equal volumes.

Add excess ammonia solution. If the precipitate dissolves, giving a deep blue solution, this confirms that Cu^{2+} ions are present.

To one, add double its volume of sodium hydroxide solution.

To the other, add double its volume of ammonium hydroxide solution.

If the precipitate …	it indicates the presence of …
dissolves in both sodium hydroxide and ammonium hydroxide solutions, giving a colourless solution	Zn^{2+} ions
dissolves in neither	Ca^{2+} ions
dissolves only in the sodium hydroxide solution, giving a colourless solution	Al^{3+} ions

✓

Quick check for 14.2   *(Answers on page 168)*

1   To test if a solution contains Cu^{2+} ions, you add sodium hydroxide solution. If Cu^{2+} ions are present, what will you see? How can you *confirm* their presence?

2   You add some sodium hydroxide to a solution that contains an unknown metal ion. A white precipitate forms. This will not dissolve in excess sodium hydroxide or ammonium hydroxide. What was the ion?

14.3 Testing for anions

Testing for halide ions (Cl⁻, Br⁻, I⁻)

By forming a precipitate. Silver halides are insoluble. So add silver nitrate, under acidic conditions, to see if a precipitate forms.

1 Take a small amount of the solution. Add an equal volume of dilute nitric acid.
2 Then add silver nitrate solution.
3 If a halide ion is present, a precipitate will form, as shown in this table:

Precipitate	Indicates presence of ...	Reaction taking place
white	chloride ions, Cl⁻	$Ag^+ (aq) + Cl^- (aq) \longrightarrow AgCl (s)$
cream	bromide ions, Br⁻	$Ag^+ (aq) + Br^- (aq) \longrightarrow AgBr (s)$
yellow	iodide ions, I⁻	$Ag^+ (aq) + I^- (aq) \longrightarrow AgI (s)$

Note that iodide ions can also be identified using acidified lead(II) nitrate solution. A deep yellow precipitate of lead(II) iodide forms.

Testing for sulfate ions (SO₄²⁻)

By forming a precipitate. Barium sulfate is insoluble. So add barium nitrate to the unknown solution, under acidic conditions. If a precipitate forms, the sulfate ion is present.

1 Take a small amount of the solution. Add an equal volume of dilute hydrochloric acid.
2 Then add barium nitrate solution.
3 If sulfate ions are present, a white precipitate will form. The reaction is:
   $Ba^{2+} (aq) + SO_4^{2-} (aq) \longrightarrow BaSO_4 (s)$

Testing for carbonate ions (CO₃²⁻)

By releasing a gas. Dilute acids react with carbonates, giving off carbon dioxide gas.

1 Take a small amount of the unknown solid or solution, and add a little dilute hydrochloric acid.
2 If the mixture bubbles and gives off a gas that turns limewater milky, the unknown substance contained carbonate ions. The reaction is:

$$2H^+ (aq) \quad + \quad CO_3^{2-} (aq) \quad \longrightarrow \quad CO_2 (g) \quad + \quad H_2O (l)$$
from the acid   (turns limewater milky)

Testing for nitrate ions (NO₃⁻)

By releasing a gas. Aluminium metal reacts with nitrates, under alkaline conditions, to release ammonia gas.

1 Take a small amount of the unknown solid or solution. Add a little sodium hydroxide solution and some aluminium foil. Heat gently.
2 If ammonia gas is given off, the unknown substance contained nitrate ions, which are being **reduced** to ammonia. The reaction that takes place is:
   $8Al (s) + 3NO_3^- (aq) + 5OH^- (aq) + 2H_2O (l) \longrightarrow 3NH_3 (g) + 8AlO_2^- (aq)$

✓

Quick check for 14.3   *(Answers on page 168)*

1 You want to test a solution to see if it contains bromide ions.
   a Which reagent will you use?
   b You see something that proves bromide ions are present. What do you see?
2 Describe how you would test for the present of sulfate ions in a compound.
3 You think a sample of rock contains carbonates. How will you test it?
4 a What would you add, to release ammonia from sodium nitrate?
   b What do you add to release ammonia, when testing for the ammonium ion?

> **Remember**
> Anions (−) would go to the anode (+) during electrolysis.

14.4 Identifying gases

Many gases are colourless. However, all have chemical properties that you can test for. (No need to rely on smell!)

Gas	Description and test details
Ammonia, NH$_3$ Properties Test Result	Ammonia is a colourless alkaline gas. Hold damp red litmus paper in it. The litmus paper turns blue.
Carbon dioxide, CO$_2$ Properties Test Result	Carbon dioxide is a colourless, weakly acidic gas. It reacts with limewater (a solution of calcium hydroxide in water), to give a white precipitate of calcium carbonate: $CO_2 (g) + Ca(OH)_2 (aq) \longrightarrow CaCO_3 (s) + H_2O (l)$ Bubble the gas through limewater. The limewater turns cloudy or milky.
Chlorine, Cl$_2$ Properties Test Result	Chlorine is a green poisonous gas that bleaches dyes. Hold damp litmus paper in the gas, in a fume cupboard. The litmus paper is bleached white.
Hydrogen, H$_2$ Properties Test Result	Hydrogen is a colourless gas that combines violently with oxygen when lit. Collect the gas in a tube and hold a lighted splint to it. The gas burns with a squeaky pop.
Oxygen, O$_2$ Properties Test Result	Oxygen is a colourless gas. Fuels burn much more readily in pure oxygen than in air. Collect the gas in a test tube and hold a glowing splint to it. The splint immediately bursts into flame.

✓ **Quick check for 14.4**   (*Answers on page 168*)

1   Of the gases above, only one is coloured. Which one?
2   On which property does the test for oxygen rely?
3   Why must chlorine be tested in a fume cupboard?
4   Which gas is flammable?
5   Ammonia has a strong sharp smell. This can give you a clue that a gas is ammonia. But how would you *confirm* that it is ammonia?
6   Explain *why* carbon dioxide turns limewater milky.

Questions on section 14

Answers for these questions are on page 168.

Core curriculum

1 A precipitate may be formed when two aqueous solutions are mixed. The colour of these precipitates may be used to identify particular aqueous ions.

    a Complete the following table.

Ion under test	Test for the ion	Colour of precipitate
iron(II)		
iodide		
chloride		
sulfate		
copper(II)		

    b How would the result of the test be different if iron(II) ions were replaced with iron(III) ions? *CIE 0620 June '03 Paper 2 Q5*

2 The following compounds dissolve in water to give colourless solutions:

A   ammonium carbonate

B   calcium nitrate

C   zinc nitrate

Chemical tests would be able to identify each solution.

    a Why would adding hydrochloric acid identify A from B and C?

    b Describe the test for the anion in B and C.

    c Describe how you would distinguish between the two cations in B and C.

    d Which of the chemicals A, B and C release ammonia gas when heated with aqueous sodium hydroxide?

3 The label on a bottle of **mineral water** lists the concentration of ions dissolved in the water in milligrams per litre.

Concentration of ions in milligrams per litre			
calcium	32	nitrate	1
chloride	5	potassium	0.5
hydrogencarbonate	133	sodium	4.5
magnesium	8	sulfate	7

    a State the name of two negative ions which appear in this list.

    b Which metal ion in this list is present in the highest concentration?

    c Calculate the amount of magnesium ions in 5 litres of this mineral water.

    d Which ion in the list reacts with aqueous silver nitrate to give a white precipitate?

    e Which ion in the list gives off ammonia when warmed with sodium hydroxide and aluminium foil?

The pH of the mineral water is 7.8.

    f Which one of the following best describes this pH?

       slightly acidic   slightly alkaline   neutral   very acidic   very alkaline

                                       CIE 0620 November '07 Paper 2 Q2

Alternative to practical

1 Three different liquids P, Q and R were analysed. Q was an aqueous solution of sodium hydroxide. The tests on the liquids and some of the observations are in the following table. Complete the missing observations in the table.

Tests	Observations			
a Test the pH of the liquids using indicator paper. Note the colour of the paper.		P	Q	R
	colour of paper	red	....	orange
	pH	1	....	5
b i Add a 5 cm piece of magnesium to about 3 cm^3 of liquid P in a test-tube. Test the gas given off.	bubbles of gas lighted splint pops			
ii Repeat **b i** using liquids Q and R. Do not test for any gases.	Q .... R			
c To about 2 cm^3 of liquid P add 1 spatula measure of sodium carbonate. Test the gas given off.	....			
d By using a teat pipette add aqueous silver nitrate to about 1 cm^3 of liquid P.	white precipitate			
e By using a teat pipette add liquid Q to about 1 cm^3 of aqueous iron(II) sulfate.	....			

f Name the gas given off in test **b i**.
g Name the gas given off in test **c**.
h Identify liquid P.
i What conclusions can you draw about liquid R?

CIE 0620 November '07 Paper 6 Q5

2 Two different solids, T and V, were analysed. T was a calcium salt.
The tests on the solids and some of the observations are in the following table.
Complete the missing observations in the table.

Tests	Observations
Tests on solid T **a** Appearance of solid T	white solid
b A little of solid T was dissolved in distilled water. The solution was divided into three test-tubes.	
i The pH of the first portion of the solution was tested.	orange colour, pH 5
ii To the second portion of the solution was added excess aqueous sodium hydroxide.	....
iii To the third portion was added excess ammonia solution.	....
Tests on solid V **c** Appearance of solid	green crystals
d A little of solid V was dissolved in distilled water. The solution was divided into three test-tubes. The smell of the solution was noted.	smells of vinegar
i Test **b i** was repeated using the first portion of the solution.	orange colour, pH 6
ii Test **b ii** was repeated using the second portion.	pale blue precipitate
iii Test **b iii** was repeated using the third portion.	pale blue precipitate soluble in excess to form a dark blue solution

e What do tests **b i** and **d i** tell you about solutions T and V?
f What further conclusions can you draw about solid V?

CIE 0620 June '08 Paper 6 Q5

15 The Periodic Table

The big picture

- There are over 100 elements.
- The key difference between them is the number of protons they have. (Each element has a unique proton number.)
- The number of protons = the number of electrons, and the number of outer-shell electrons dictates an element's chemical behaviour.
- The Periodic Table shows the elements in order of increasing proton number, and at the same time groups elements with similar behaviour together.
- We can also mark in a zig-zag line separating the metals from the non-metals.

15.1 An overview of the Periodic Table

The **Periodic Table** is the chemist's way of classifying the elements.
Take a look at the Periodic Table on the next page first, before you read on.

The groups

- The elements in a group have similar properties.
- They have similar properties because their atoms have the same number of **valency electrons** (or outer-shell electrons). This shows the pattern:

Group	I	II	III	IV	V	VI	VII	0
Number of valency electrons	1	2	3	4	5	6	7	8 (or 2)

- So the group number is the same as the number of valency electrons – except for Group 0, where the 0 tells you that the atoms have a stable arrangement of 8 valency electrons (or 2 for helium).
- Given the properties of one element in a group, chemists can **predict** the properties of others.

The periods

- As you go along a period from left to right, the proton numbers of the elements increase by 1 each time.
- The elements in a period *do not* have similar properties. In fact, there is a trend (gradual change) from **metal** to **non-metal** properties along the period, from left to right.
- The properties change because the number of valency electrons increases.
- Knowing this trend allows chemists to **predict** the properties of different elements in the same period.

The zig-zag line

- This separates the **metals** from the **non-metals**, with the metals on the left. (There are far more metals than non-metals.)
- The elements in Groups I and II are all metals; those in Groups VII and 0 are all non-metals.
- But note that there is a change from non-metal to metal as you go **down** the group, in Groups III to VI. (The zig-zag line goes through these groups.)
- Knowing this trend allows chemists to **predict** the properties of different elements in the same group.

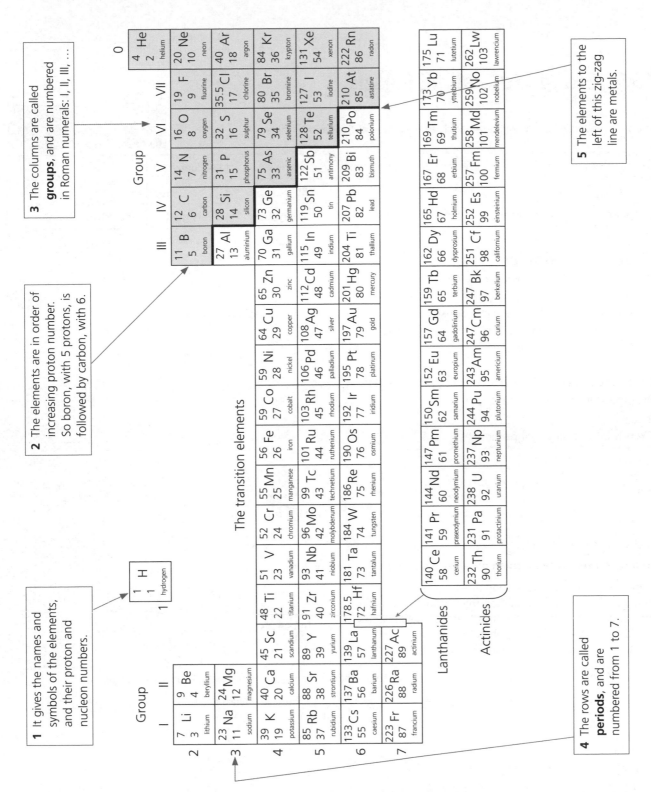

3 The columns are called **groups**, and are numbered in Roman numerals: I, II, III, …

Group

2 The elements are in order of increasing proton number. So boron, with 5 protons, is followed by carbon, with 6.

5 The elements to the left of this zig-zag line are metals.

The transition elements

1 It gives the names and symbols of the elements, and their proton and nucleon numbers.

	1 H
	1
	hydrogen

Lanthanides

Actinides

4 The rows are called **periods**, and are numbered from 1 to 7.

Group

Notice how hydrogen stands alone at the start of the table. It forms positive ions, H$^+$, like a metal, but acts as a non-metal in other ways.

✓
Quick check for 15.1   (Answers on page 168)

1 The elements in the Periodic Table are arranged in order of their ………… ?

2 Which have similar properties?
  A elements from the same group
  B elements from the same period

3 How does the number of valency electrons change, across a period?

4 Which groups contain *only* non-metals?

5 Using only the Periodic Table above, describe how the character of the elements changes in going down Group IV.

15.2 Group I: the alkali metals

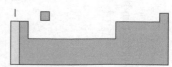

The six members of the alkali metal family form Group I of the Periodic Table.

Alkali metal	lithium	sodium	potassium	rubidium	caesium	francium
Symbol	Li	Na	K	Rb	Cs	Fr
Appearance	silvery white	silvery white	silvery white	silvery grey	silvery gold	*radioactive, very rare*

They are all similar

They are all **metals**, but compared to most other metals:

*   they are much softer (all can be cut with a knife)
*   they are lighter – they have low **density** (lithium, sodium and potassium float on water)
*   they have much lower melting and boiling points
*   they are much more reactive.

But they are not all exactly the same

The properties change down the group.

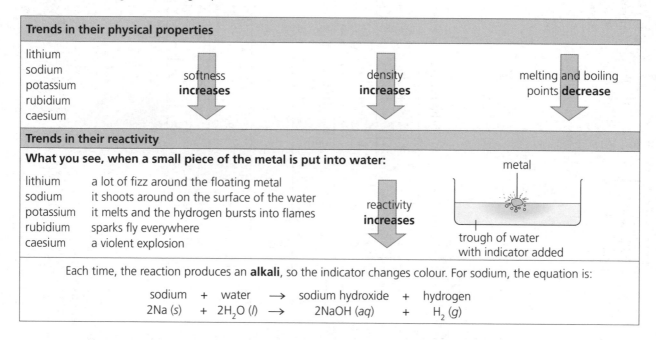

Trends in their physical properties

lithium
sodium
potassium
rubidium
caesium

softness **increases**

density **increases**

melting and boiling points **decrease**

Trends in their reactivity

What you see, when a small piece of the metal is put into water:

lithium	a lot of fizz around the floating metal
sodium	it shoots around on the surface of the water
potassium	it melts and the hydrogen bursts into flames
rubidium	sparks fly everywhere
caesium	a violent explosion

reactivity **increases**

metal

trough of water
with indicator added

Each time, the reaction produces an **alkali**, so the indicator changes colour. For sodium, the equation is:

sodium + water ⟶ sodium hydroxide + hydrogen

$$2Na\,(s) + 2H_2O\,(l) \longrightarrow 2NaOH\,(aq) + H_2\,(g)$$

✓

Quick check for 15.2 *(Answers on page 168)*

1   Which group of the Periodic Table contains the alkali metals?

2   Which alkali metal has the highest boiling point?

3   Which is softer, sodium or potassium?

4   Describe the trend in reactivity as you go *up* the alkali metal group.

5   **Francium** is a very rare and unstable element. Its properties have not been established by experiment.

   a What physical properties would you predict, for francium?

   b Which other alkali metal would francium be most like, in its reactivity?

   c How would francium compare to this other alkali metal?

Extended

15.3 Group VII: the halogens

The five members of the halogen family make up Group VII of the Periodic Table.

Halogen	fluorine	chlorine	bromine	iodine	astatine
Symbol	F	Cl	Br	I	At
Appearance	yellow gas	green gas	dark red liquid	black solid	*radioactive, very rare*

They are all similar

- They are all poisonous non-metals, which form coloured gases.
- They exist as **diatomic** molecules (two atoms in each molecule). For example, Cl_2.
- They are more reactive than most other non-metals.

But they are not all exactly the same

The properties change down the group.

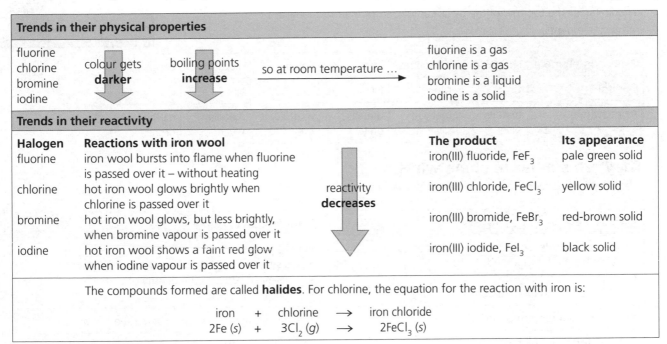

Trends in their physical properties		
fluorine chlorine bromine iodine	colour gets **darker** → boiling points **increase** → so at room temperature …	fluorine is a gas chlorine is a gas bromine is a liquid iodine is a solid

Trends in their reactivity

Halogen	Reactions with iron wool		The product	Its appearance
fluorine	iron wool bursts into flame when fluorine is passed over it – without heating		iron(III) fluoride, FeF_3	pale green solid
chlorine	hot iron wool glows brightly when chlorine is passed over it	reactivity **decreases**	iron(III) chloride, $FeCl_3$	yellow solid
bromine	hot iron wool glows, but less brightly, when bromine vapour is passed over it		iron(III) bromide, $FeBr_3$	red-brown solid
iodine	hot iron wool shows a faint red glow when iodine vapour is passed over it		iron(III) iodide, FeI_3	black solid

The compounds formed are called **halides**. For chlorine, the equation for the reaction with iron is:

$$\text{iron} + \text{chlorine} \longrightarrow \text{iron chloride}$$
$$2Fe\,(s) + 3Cl_2\,(g) \longrightarrow 2FeCl_3\,(s)$$

Reactions of the halogens with halide ions

Compare these reactions of the halogens with solutions containing other **halide ions**.
(The solutions are potassium halides dissolved in water.)

If the solution contains …	chloride ions (Cl^-)	bromide ions (Br^-)	iodide ions (I^-)
when chlorine (Cl_2) is added		the solution turns orange	the solution turns red-brown
when bromine (Br_2) is added	there is no change		the solution turns red-brown
when iodine (I_2) is added	there is no change	there is no change	

These reactions show that, of the three halogens:
- chlorine is the **most** reactive; it **displaces** both bromine and iodine from their compounds.
- iodine is the **least** reactive; it does not displace either of the others from their compounds.

This shows how chlorine displaces bromine:

$$Cl_2\,(g) + 2KBr\,(aq) \longrightarrow 2\,KCl\,(aq) + Br_2\,(aq)$$
$$\text{colourless} \qquad\qquad\qquad \text{orange}$$

The ionic equation is: $Cl_2\,(g) + 2Br^-\,(aq) \longrightarrow 2Cl^-\,(aq) + Br_2\,(aq)$

Remember
An ionic equation leaves out ions that do not take part in the reaction.

15.4 The transition elements

There are thirty transition elements. They form a block in the middle of the Periodic Table. The top row of the block contains six of the most familiar transition elements:

The transition elements

Transition element	chromium	manganese	iron	cobalt	nickel	copper
Symbol	Cr	Mn	Fe	Co	Ni	Cu
Appearance	metallic silver	metallic silver	metallic grey	metallic grey	metallic silver	metallic bronze

They are similar in some ways

The transition elements have some general similarities:

Their similarities	Using iron as example
They are all **metals**, and share these properties: • they are hard, tough and strong • they have high density • they have high melting points.	Iron: • is strong enough to be used in building bridges • has nearly 8 times the density of water • melts at 1535 °C.
They form coloured compounds.	The compounds of iron are usually green or brown.
They can form ions with different charges – in other words, they have **variable valency**. (The different oxidation states lead to the different colours of compound.)	Iron can form Fe^{2+} and Fe^{3+} ions. The name of the iron compound tells you which ion is in it: iron(II) oxide contains Fe^{2+} ions iron(III) oxide contains Fe^{3+} ions.
The elements and their compounds often act as **catalysts** for other reactions. (See page 68.)	Iron is used as a catalyst in the manufacture of ammonia. (See page 133.)

15.5 Group 0: the noble gases

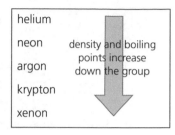

Six elements form the noble gas family, which is Group 0 of the Periodic Table.

Noble gas	helium	neon	argon	krypton	xenon	radon
Symbol	He	Ne	Ar	Kr	Xe	Rn
Appearance	colourless gas	colourless gas	colourless gas	colourless gas	colourless gas	colourless gas (*radioactive*)

Similar – but not exactly the same
- The Group 0 elements are all non-metals.
- They are colourless gases, and occur naturally in air.
- They exist as single atoms – they are **monatomic**.
- They are **unreactive** or **inert** – they do not normally react with anything.
- But their physical properties show trends down the group, as shown on the right.

helium

neon

argon — density and boiling points increase down the group

krypton

xenon

Why are they unreactive?
The noble gases are unreactive because their atoms have a very stable arrangement of outer-shell electrons. They do not need to bond to other atoms to gain, or lose, electrons.

Their uses
The unreactivity of the noble gases makes them very useful. For example:
- helium is used in balloons and airships because it is lighter than air, and non-flammable
- argon is used to provide an inert atmosphere, for example in welding and in tungsten light bulbs. (Oxygen would react with hot metals being welded, and hot tungsten.)

✓
Quick check for 15.5   (*Answers on page 168*)
1   All the noble gases are *unreactive*, and *monatomic*. Why is this?
2   Give two reasons why helium is used in airships.

15.6 Across a period

Extended

The elements change from metal to non-metal, across a period. Look at Period 3.

Group	I	II	III	IV	V	VI	VII	0
Element	sodium	magnesium	aluminium	silicon	phosphorus	sulfur	chlorine	argon
No. of valency electrons	1	2	3	4	5	6	7	8 (stable arrangement)
Element is a …	metal	metal	metal	metalloid	non-metal	non-metal	non-metal	non-metal
Oxide is …	basic	basic	amphoteric	acidic	acidic	acidic	acidic	———

Silicon is called a **metalloid** because it acts as a metal in some ways, and as a non-metal in others. And note the change in the oxides across the period.

✓
Quick check for 15.6   (*Answers on page 168*)
1   How do the oxides of the elements change across a period?
2   Aluminium forms an *amphoteric* oxide. What does that mean? (Check page 89.)

Questions on section 15

Answers for these questions are on page 168.

Core curriculum

1 The table below shows an early form of the Periodic Table made by John Newlands in 1866:

H	F	Cl	Co, Ni	Br
Li	Na	K	Cu	Rb
Be	Mg	Ca	Zn	Sr
B	Al	Cr	Y	
C	Si	Tl	In	
N	P	Mn	As	
O	S	Fe	Sc	

a Newlands arranged the elements according to their relative atomic masses. What governs the order of the elements in the modern Periodic Table?

b Use your modern Periodic Table to suggest why Newlands put cobalt and nickel in the same place.

c Which group of elements is missing from Newlands' table?

d Describe three other differences between Newlands' table and the modern Periodic Table. You must not give any of the answers you mentioned in parts **a**, **b**, or **c**.

CIE 0620 November '07 Paper 2 Q6

2 The table below gives information about the elements in Group I of the Periodic Table.

Element	Boiling point / °C	Density / g cm^{-3}	Radius of the atom in the metal / nm	Reactivity with water
lithium	1342	0.53	0.157	
sodium	883	0.97	0.191	rapid
potassium	760	0.86	0.235	very rapid
rubidium		1.53	0.250	extremely rapid
caesium	669	1.88		explosive

a How does the density of the Group I elements change down the group?

b Suggest a value for the boiling point of rubidium.

c Suggest a value for the radius of a caesium atom.

d Use the information in the table to suggest how fast lithium reacts with water compared with the other Group I metals.

e State three properties shown by **all** metals. *CIE 0620 November '04 Paper 2 Q1*

3 Chlorine is in Group VII of the Periodic Table. When chlorine reacts with a solution of potassium iodide, the solution turns a reddish-brown colour.

a Write a word equation for this reaction.

b Explain why iodine does not react with a solution of potassium bromide.

CIE 0620 June '08 Paper 2 Q4

4 Pure air contains about 1% argon.

a In which period of the Periodic Table is argon?

b State the name of the group of elements to which argon belongs.

c Draw the electronic structure of argon.

d Why is argon used in lamps? *CIE 0620 June '07 Paper 2 Q2*

5 The halogens are a group of elements showing trends in colour, state and reaction with other halide ions.

a Write the word equation for the reaction of chlorine with aqueous potassium bromide.

b Explain why an aqueous solution of iodine does not react with potassium chloride.

The table shows the properties of some halogens.

Halogen	State at room temperature	Colour	Boiling point / °C	Density of solid / g cm^{-3}
fluorine	gas	yellow		1.51
chlorine		green	−35	1.56
bromine	liquid	red-brown	59	
iodine	solid		184	4.93

c **i** What is the state of chlorine at room temperature?

   ii What is the colour of iodine?

d Suggest values for:

   i the boiling point of fluorine,

   ii the density of bromine.

e How many electrons does an atom of fluorine have

   i in total,

   ii in its outer shell? *CIE 0620 June '07 Paper 2 Q6*

Extended curriculum

1 For each of the following select an element from Period 4, potassium to krypton, that matches the description.

a It is a brown liquid at room temperature.

b It forms a compound with hydrogen having the formula XH_4.

c A metal that reacts violently with cold water.

d It has a complete outer energy level.

e It has oxidation states of 2 and 3 only.

f It can form an ion of the type X^-. *CIE 0620 June '08 Paper 3 Q1*

2 Use the Periodic Table on page 105 to help you answer these questions.

a Predict the formula of each of the following compounds.

   i barium oxide

   ii boron oxide

b Give the formula of the following ions.

   i sulfide

   ii gallium

c Draw a diagram showing the arrangement of the valency electrons in one molecule of the covalent compound nitrogen trichloride.

   Use × to represent an electron from a nitrogen atom.

   Use o to represent an electron from a chlorine atom.

d Potassium and vanadium are elements in Period IV.

   i State two differences in their physical properties.

   ii Give two differences in their chemical properties.

e Fluorine and astatine are halogens. Use your knowledge of the other halogens to predict the following:

   i the physical state of fluorine at room temperature and pressure (rtp)

   ii the physical state of astatine at rtp

   iii two similarities in their chemical properties. *CIE 0620 June '07 Paper 3 Q4*

Extended

16 Metals

The big picture

- Metals are all different, but do share a number of properties we find useful.
- Often we add other substances to them to make them more useful – for example, to make them harder and stronger.
- If a metal is very reactive, it means it has a very strong drive to exist as positive ions. So its compounds are stable, and hard to break down.
- We can carry out experiments to rank metals in order of reactivity.
- Most metals occur as compounds in their ores, and we have to extract them. The method depends on their reactivity.

16.1 Properties of metals

- All metals are good conductors of **electricity** and **heat.**
- Metals share many of these **physical** properties:

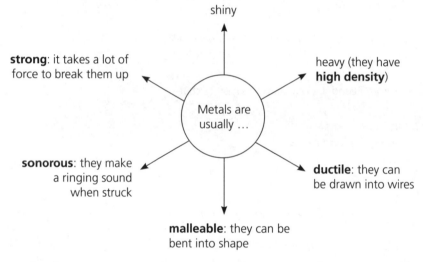

But there are many exceptions. For example mercury is a liquid at room temperature, aluminium is a light metal, and sodium is very soft.

- Most metals undergo these two **chemical** reactions:

- When metals react, they lose electrons and form **positive ions**.

✓

Quick check for 16.1   (*Answers on page 168*)

1 What does this term mean?   **a** malleable   **b** sonorous   **c** ductile
2 Give two ways in which most metals are similar in their chemical reactions.
3 Name the compound formed, and give the symbol for its metal ion, when:
   a calcium reacts with oxygen   **b** aluminium reacts with hydrochloric acid

16.2 Alloys

Often, pure metals are not up to the job! They may be too soft, or easily corroded by air and water. We may be able to solve the problem by making an **alloy**.

An alloy is a **mixture** where at least one other substance is added to a metal, to improve on its properties. The other substance is often – but not always – another metal.

Look at these diagrams:

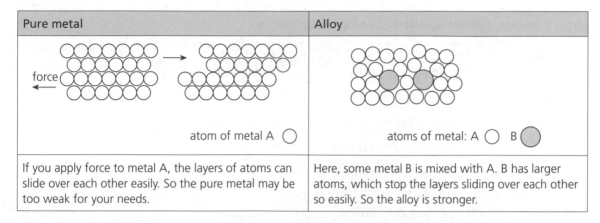

Pure metal	Alloy
atom of metal A ○	atoms of metal: A ○ B ●
If you apply force to metal A, the layers of atoms can slide over each other easily. So the pure metal may be too weak for your needs.	Here, some metal B is mixed with A. B has larger atoms, which stop the layers sliding over each other so easily. So the alloy is stronger.

Examples of alloys

Alloys are designed to meet different needs. Modern alloys may contain small amounts of several substances, as in the aluminium example below.

Alloy	Made from	Special properties	Used for ...
stainless steel	70% iron 20% chromium 10% nickel	does not rust	car parts, kitchen sinks, cutlery, tanks and pipes in factories, surgical instruments
aluminium alloy 7075 TF	90.25% aluminium 6% zinc 2.5% magnesium 1.25% copper	light but very strong	aircraft
brass	70% copper 30% zinc	harder than copper; does not corrode	musical instruments, ornaments, door knobs and other fittings
bronze	95% copper 5% tin	harder than brass, does not corrode, chimes when struck	statues, ornaments, church bells

✓

Quick check for 16.2   *(Answers on page 168)*

1   Pure iron would not be suitable for making car parts. Why not?

2   How can the properties of a metal be improved on?

3   **a** What is brass?

      b Its properties are different from those of pure copper. Give examples.

4   Using a diagram, show why an alloy is usually stronger than the pure metal.

16.3 The reactivity series for metals

Comparing reactivity

Metals differ in how **reactive** they are. We can carry out experiments to put them in order of reactivity. These are suitable reactions:

- reactions of the **metals** with water or steam, and with dilute hydrochloric acid
- reactions of the **metal oxides** with carbon (to try to reduce them).

Reactions of the metals

The more vigorous a reaction, the more reactive the metal is. Look at the order:

Order of reactivity	Metal	Reaction with water or steam	Reaction with hydrochloric acid
most reactive	potassium	very violent with cold water	too dangerous to try
	sodium	violent with cold water [1]	too dangerous to try
	calcium	less violent with cold water	very vigorous
	magnesium	very slow with cold water vigorous with steam [2]	vigorous
	zinc	quite slow with steam	moderate [3]
	iron	slow with steam	slow
least reactive	copper	no reaction	no reaction

Equations for some of the reactions

These are the equations for the numbered reactions in the table above:

1   $2Na\ (s) + 2H_2O\ (l) \longrightarrow 2NaOH\ (aq) + H_2\ (g)$

2   $Mg\ (s) + H_2O\ (g) \longrightarrow MgO\ (s) + H_2\ (g)$

3   $Zn\ (s) + 2HCl\ (aq) \longrightarrow ZnCl_2\ (aq) + H_2\ (g)$

Note that:

- if the metals react with cold water, they form **hydroxides**
- those that react with steam form **oxides**
- and when a metal reacts with cold water, steam, or acid, **hydrogen** always forms.

Reactions of the metal oxides

Carbon will remove oxygen from an oxide *only* if the metal is *less reactive* than carbon. Loss of oxygen is **reduction**. So carbon acts as a **reducing agent**. Look at the results:

Order of reactivity	Metal	Reaction of metal oxide on heating with carbon
most reactive	potassium	no reaction
	sodium	no reaction
	calcium	no reaction
	magnesium	no reaction
	zinc	reduction to zinc takes place, but only at a very high temperature [2]
	iron	reduction to iron
least reactive	copper	easy reduction to copper [1]

Equations for some of the reactions
These are the equations for the numbered reactions in the last table above:

1 $2CuO\,(s) + C\,(s) \longrightarrow 2Cu\,(s) + CO_2\,(g)$

2 $ZnO\,(s) + C\,(s) \longrightarrow Zn\,(s) + CO\,(g)$

Each time, carbon brings about reduction, and is itself oxidised.
It can be oxidised to either carbon dioxide (CO_2) or carbon monoxide (CO).

Putting it all together: the reactivity series
Combining the results of those experiments gives the reactivity series:

The reactivity series		
potassium, K sodium, Na calcium, Ca magnesium, Mg aluminium, Al	most reactive ↑	*carbon can't reduce the oxides of metals above this line* ↘
carbon zinc, Zn iron, Fe lead, Pb	increasing reactivity	*metals above this line displace hydrogen from acids* ↘
hydrogen copper, Cu silver, Ag gold, Au	least reactive	

Note that:
- the most reactive metals in the series belong to Groups I and II of the Periodic Table.
- the less reactive metals, from zinc downwards, are transition metals (except for lead).
- we could add more metals to the series (for example tin, chromium, nickel) by carrying out experiments with those metals too.

The reactivity of aluminium
Notice that aluminium is quite high in the reactivity series. But in everyday use, it appears unreactive. (For example we use it in window frames.)

In fact aluminium reacts quickly with oxygen, forming a thin layer of aluminium oxide on its surface. But this layer then sticks to the surface and stops further oxygen getting through.

✓ **Quick check for 16.3**   *(Answers on page 168)*
1 Which gas is released when a metal reacts with water?
2 **a** Describe what you would observe when magnesium reacts with cold water.
   b Write the equation for this reaction.
   c What would happen if you replaced the water by steam?
3 No experiments were carried out for the reactions of potassium and sodium with hydrochloric acid. Why not?
4 Carbon is not a metal. But it is included in the reactivity series above. Explain why.
5 List the transition metals in the series above, and comment on their reactivity compared with Group I metals.
6 What is unusual about the reactivity of aluminium?

Extended

Extended

16.4 More about the reactivity series

The reactivity series and competition reactions

The reactivity series shows metals in order of their reactivity. But what this really means is:
it shows metals in order of their tendency to form positive ions.

We can show this by reacting a metal with the compound of another metal. The two metals
then compete to be ions, and the more reactive one 'wins'. Compare these reactions:

	Competing for oxygen	Competing to form ions in solution
Reaction	When iron is heated with copper(II) oxide, an exothermic reaction takes place. Iron(II) oxide forms, and bronze-coloured copper appears.	When an iron nail is stood in blue copper(II) sulfate solution, a coating of copper forms on the nail, and the solution turns green.
Explanation	Iron has taken the oxygen from the copper, because iron has a stronger tendency than copper to exist as positive ions.	Iron ions have replaced the copper ions in solution, because iron has a stronger tendency than copper to exist as positive ions.
Equation	$Fe\ (s)\ +\ CuO\ (s) \longrightarrow FeO\ (s)\ +\ Cu\ (s)$	$Fe\ (s)\ +\ CuSO_4\ (aq) \longrightarrow FeSO_4\ (aq)\ +\ Cu\ (s)$
The two half equations	$Fe \longrightarrow Fe^{2+}\ +\ 2e^-$ (shows electron loss) $Cu^{2+}\ +\ 2e^- \longrightarrow Cu$ (shows electron gain)	$Fe \longrightarrow Fe^{2+}\ +\ 2e^-$ $Cu^{2+}\ +\ 2e^- \longrightarrow Cu$
The ionic equation	$Fe\ +\ Cu^{2+} \longrightarrow Fe^{2+}\ +\ Cu$ (add the half equations, and cancel the electrons)	$Fe\ +\ Cu^{2+} \longrightarrow Fe^{2+}\ +\ Cu$
Conclusion	In each case, iron has given up electrons to form positive ions. **The more reactive metal forms positive ions more readily.**	

The reactivity series and the stability of compounds

The more reactive a metal, the better it is at 'hanging on' to other elements in its
compounds. So the more energy is needed to break the compounds down.
Compare what happens when compounds of sodium and copper are heated:

Compounds	Effect of heat on the sodium compounds	Effect of heat on the copper compounds
carbonate	No reaction; the white solid remains unchanged.	The green compound breaks down easily, giving black copper(II) oxide: $CuCO_3\ (s) \xrightarrow{heat} CuO\ (s)\ +\ CO_2\ (g)$
hydroxide	No reaction; the white solid remains unchanged.	The pale blue compound breaks down easily, giving black copper(II) oxide: $Cu(OH)_2\ (s) \xrightarrow{heat} CuO\ (s)\ +\ H_2O\ (l)$
nitrate	Partial decomposition occurs. Oxygen is given off, and a nitrite forms: $2NaNO_3\ (s) \xrightarrow{heat} 2NaNO_2\ (s)\ +\ O_2\ (g)$ sodium nitrite	The bright blue compound breaks down easily, giving black copper(II) oxide. A brown gas also forms. It is nitrogen dioxide, NO_2: $2Cu(NO_3)_2\ (s) \xrightarrow{heat} 2CuO\ (s)\ +\ 4NO_2\ (g)\ +\ O_2\ (g)$
	So the compounds of the less reactive metal break down more easily.	

The general rules for thermal decomposition

The breaking down of a compound by heat is called **thermal decomposition**.
- The lower a metal is in the reactivity series, the more readily its compounds decompose when heated**.**
- Carbonates *except sodium and potassium* decompose to the oxide and carbon dioxide.
- Hydroxides *except sodium and potassium* decompose to the oxide and water.
- Nitrates *except sodium and potassium* decompose to the oxide, nitrogen dioxide and water.

✓

Quick check for 16.4 *(Answers on page 168)*

1 Iron powder is added to blue copper(II) sulfate solution. The solution goes green. Explain why, and give the word equation.

2 **a** Write the equation for the reaction between magnesium and lead(II) oxide.
  b Now write the two half equations for the reaction. (See pages 79 and 80.)
  c Add the balanced half equations, to give the ionic equation for the reaction.

3 Write an equation for the thermal decomposition of:
  a potassium nitrate **b** iron(III) hydroxide **c** calcium carbonate.

> **Remember**
> If metal X pushes metal Y out of compounds, and takes its place, then metal X is more reactive than metal Y.

16.5 Extraction of metals

A rock that contains enough metal to make it worth mining is called an **ore**.
Only the most unreactive metals, such as gold and silver, occur as **elements** in their ores.
The rest are found as **compounds**, from which the metals have to be extracted.

Methods of extraction

- The more reactive a metal, the better it is at 'hanging on' to the elements in its compounds.
- So the more reactive the metal, the more difficult it is to extract.
- So the method of extraction is linked to the metal's position in the **reactivity series**.

Metal			Method of extraction from ore		
potassium			electrolysis – the most powerful method of extraction; but it uses a lot of electricity so it is the most expensive method too		
sodium					
calcium					
magnesium					
aluminium	more reactive	more difficult to extract		more powerful extraction method	more expensive process
carbon					
zinc			heat with a reducing agent: carbon or carbon monoxide		
iron					
lead					
silver			occur naturally as elements, so no chemical change needed		
gold					

Extraction is reduction

The chemical change to obtain a metal from its ore is always a **reduction** reaction.
For example aluminium is obtained by extracting it from aluminium oxide, which is made from its ore **bauxite**. The extraction is carried out using electrolysis:

$$2Al_2O_3 \ (l) \xrightarrow{\text{electricity}} 4Al \ (l) \ + \ 3O_2 \ (g) \qquad \text{(reduction is loss of oxygen)}$$

The half equation is:
$$Al^{3+} + 3e^- \longrightarrow Al \qquad \text{(reduction is gain of electrons)}$$

Either way, it is clear that the aluminium is reduced.

> **Make the link to ...**
> the extraction of aluminium, on page 53.

✓

Quick check for 16.5 *(Answers on page 168)*

1 Most metals are found as compounds in their ores. Explain why.

2 Why is aluminium extracted by electrolysis?

16.6 Extracting iron and zinc

Iron and zinc are both extracted using the gas carbon monoxide.

Extracting iron

Iron is by far the most widely used of all metals. It is extracted by reducing iron(III) oxide, using **carbon monoxide**, in a **blast furnace**. There are three **raw materials**:

- **haematite** – the iron ore; it is mainly iron(III) oxide mixed with sand (silicon dioxide)
- **coke** – almost pure carbon
- **limestone** – calcium carbonate, to remove the sand.

Air is also needed: it is blasted in at the bottom of the furnace.

The stages in the process	The chemical equations
1 Coke burns in oxygen from the air, to give carbon dioxide.	$C\ (s)\ +\ O_2\ (g)\ \longrightarrow\ CO_2\ (g)$
2 More coke **reduces** the carbon dioxide to carbon monoxide.	$C\ (s)\ +\ CO_2\ (g)\ \longrightarrow\ 2CO\ (g)$
3 The iron ore is **reduced** to iron by the carbon monoxide.	$Fe_2O_3\ (s)\ +\ 3CO\ (g)\ \longrightarrow\ 2Fe\ (s)\ +\ CO_2\ (g)$
4 Removing the sand: limestone breaks down to calcium oxide, a basic oxide. This reacts with silicon dioxide (acidic) forming **slag** (calcium silicate).	$CaO\ (s)\ +\ SiO_2\ (s)\ \longrightarrow\ CaSiO_3\ (s)$

Note: this is the crucial step.

There are two **waste gases**:

- **nitrogen** – the main gas in air, which has not reacted
- **carbon dioxide**– from the reduction reaction in stage 3.

Extracting zinc

Zinc is extracted by reducing zinc oxide using **carbon monoxide**, in a **smelting furnace**. There are three raw materials:

- **zinc blende** – zinc ore (mainly zinc sulfide) mixed with sand (silicon dioxide)
- **coke** – almost pure carbon
- **limestone** – calcium carbonate, to remove the sand.

Air is also needed, for the roasting of the zinc sulfide.

> **Remember**
> On heating:
> calcium carbonate $\longrightarrow$ calcium oxide (a base) + carbon dioxide.

The stages in the process	The chemical equations
1 Zinc sulfide is roasted in air to give zinc oxide and sulfur dioxide.	$2ZnS\ (s)\ +\ 3O_2\ (g)\ \longrightarrow\ 2ZnO\ (s)\ +\ 3SO_2\ (g)$
2 Coke burns in oxygen from the air, to give carbon dioxide. This reacts with more coke to give carbon monoxide.	$C\ (s)\ +\ O_2\ (g)\ \longrightarrow\ CO_2\ (g)$ $C\ (s)\ +\ CO_2\ (g)\ \longrightarrow\ 2CO\ (g)$
3 The zinc oxide is **reduced** to zinc by the carbon monoxide.	$ZnO\ (s)\ +\ CO\ (g)\ \longrightarrow\ Zn\ (s)\ +\ CO_2\ (g)$
4 Removing the sand: limestone breaks down to calcium oxide, a basic oxide. This reacts with silicon dioxide (acidic) to form **slag** (calcium silicate).	$CaO\ (s)\ +\ SiO_2\ (s)\ \longrightarrow\ CaSiO_3\ (s)$

Note: this is the crucial step.

There are three **waste gases**:

- **sulfur dioxide** – from the roasting of the ore
- **nitrogen** – the main gas in air, which has not reacted
- **carbon dioxide** – from the reduction reaction in stage 3.

> **Note**
> The slag from the blast furnace is used in making concrete, and building roads.

Extended

✓ **Quick check for 16.6**   (*Answers on page 168*)

1   Write a word equation for the crucial step in the extraction of iron.
2   What is the name and formula of the main ore of zinc?
3   What is the first step in the extraction of zinc?
4   The extractions of both zinc and iron require limestone. Why?

16.7 Making use of metals

Steels: alloys of iron

Iron from the blast furnace is called **pig iron**. It contains many impurities, such as phosphorus compounds, silicon dioxide (sand), and about 5% carbon. These make it hard but brittle – it can shatter easily. To improve on its properties, most iron in turned into alloys called **steels**.

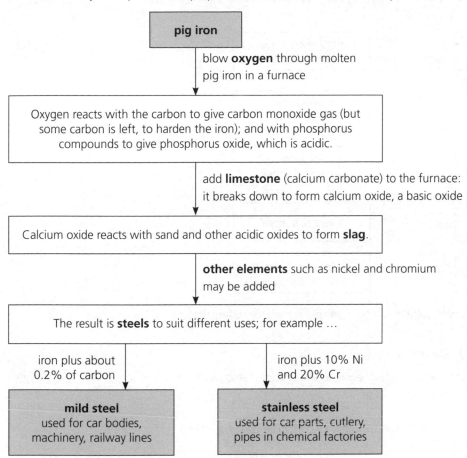

Remember

Making steels: first **remove** unwanted impurities, then **add** other elements as required.

pig iron

blow **oxygen** through molten pig iron in a furnace

Oxygen reacts with the carbon to give carbon monoxide gas (but some carbon is left, to harden the iron); and with phosphorus compounds to give phosphorus oxide, which is acidic.

add **limestone** (calcium carbonate) to the furnace: it breaks down to form calcium oxide, a basic oxide

Calcium oxide reacts with sand and other acidic oxides to form **slag**.

other elements such as nickel and chromium may be added

The result is **steels** to suit different uses; for example …

iron plus about 0.2% of carbon

iron plus 10% Ni and 20% Cr

mild steel
used for car bodies, machinery, railway lines

stainless steel
used for car parts, cutlery, pipes in chemical factories

Some uses for aluminium, zinc and copper

Metal	Use	Relevant property
aluminium	aircraft bodies	low density (light) but strong
	food containers	resists corrosion
zinc	galvanising iron	prevents the rusting of iron (page 129)
	for making brass (70% Cu, 30% Zn) for musical instruments, and door fittings	harder than copper, and is resistant to corrosion
copper	electrical wires	good conductor of electricity, and is ductile
	saucepans	good conductor of heat, and has a high melting point

Extended

✓
Quick check for 16.7   *(Answers on page 168)*

1 What is pig iron, and why is there not much demand for it?
2 How is the carbon content reduced, in the manufacture of steel?
3 Compare the compositions of mild and stainless steel.
4 Why is aluminium used in aircraft manufacture?

Questions on section 16

Answers for these questions start on page 168.

Core curriculum

1 This question is about metals.

a Match up the metals in the boxes on the left with the descriptions in the boxes on the right.

potassium	a metal used to make aircraft bodies
silver	a metal used in jewellery
aluminium	a metal extracted from haematite
platinum	a very soft metal
iron	an unreactive metal used for electrodes

b Iron powder reacts rapidly with sulfuric acid to form aqueous iron(II) sulfate and hydrogen.

$$Fe\ (s)\ +\ H_2SO_4\ (aq)\ \longrightarrow\ FeSO_4\ (aq)\ +\ H_2\ (g)$$

Describe two things that you would see happening as this reaction takes place.

c Alloys are often more useful than pure metals.

i What is an *alloy*?

ii The properties of which metal can be changed, by controlled use of additives to form **steel** alloys?

iii Does increasing the amount of carbon in a steel make it stronger, or weaker?

iv Name one other alloy apart from steel.

v Iron rusts very easily. Describe two methods of preventing rusting.

CIE 0620 November '07 Paper 2 Q3

2 Some reactions of metals W, X, Y and Z are given below.

Metal	Reaction with water	Reaction with dilute hydrochloric acid
W	A few bubbles form slowly in cold water.	Vigorous reaction. Gas given off.
X	Vigorous reaction. Metal melts. Gas given off.	Explosive reaction. Should not be attempted.
Y	No reaction.	No reaction.
Z	Does not react with cold water. Hot metal reacts with steam.	Steady fizzing.

a Arrange these metals in order of reactivity.

b Which of these metals could be: **i** magnesium, **ii** copper?

c Name the gas given off in reactions with W, X and Z.

CIE 0620 June '06 Paper 3 Q2

3 Iron is extracted from its ore in a blast furnace using carbon (coke) as a reducing agent and as a source of heat.

a The coke burns in hot air. The equation for this reaction is
$$2C(s) + O_2(g) \longrightarrow 2CO(g)$$
State the name of the gas produced in this reaction.

b Near the top of the blast furnace, the iron(III) oxide in the ore is reduced to iron.
$$Fe_2O_3(s) + 3CO(g) \longrightarrow 2Fe(l) + 3CO_2(g)$$
Use the equation to explain why this change is a reduction reaction.

c In the hottest regions of the furnace, iron(III) oxide is reduced by carbon. Complete the equation for this reaction.
$$Fe_2O_3 (s) + \text{.......}C (s) \longrightarrow \text{........}Fe (l) + 3CO (g)$$

d The iron from the blast furnace contains up to 10% by mass of impurities. The main impurities are carbon, silicon and phosphorus. The diagram below shows one method of making steel from iron.

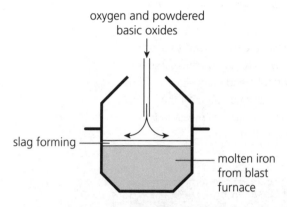

A mixture of oxygen and basic oxides is blown onto the surface of the molten iron.

i What is the purpose of blowing oxygen onto the molten iron?

ii A large amount of energy is released in the process of steelmaking. What name is given to chemical reactions which release energy?

iii The basic oxides react with the impurities in the iron and form a slag. What information in the diagram suggests that the slag is less dense than the molten iron?

iv Which one of the following is a basic oxide?
calcium oxide   carbon dioxide   sulfur dioxide   water

v Why is steel rather than iron used for constructing buildings and bridges?

CIE 0620 June '04 Paper 2 Q6

Extended curriculum

Extended

1 Titanium is produced by the reduction of its chloride. This is heated with magnesium in an inert atmosphere of argon.
$$TiCl_4 + 2Mg \longrightarrow Ti + 2MgCl_2$$
a Explain why it is necessary to use argon rather than air.

b Name another metal that would reduce titanium chloride to titanium.

c Suggest how you could separate the metal, titanium, from the soluble salt magnesium chloride.

CIE 0620 June '07 Paper 3 Q5

2 Copper has the structure of a typical metal. It has a lattice of positive ions and a 'sea' of mobile electrons. The lattice can accommodate ions of a different metal.

Give a different use of copper that depends on each of the following:

a the ability of the ions in the lattice to move past each other

b the presence of mobile electrons

c the ability to accommodate ions of a different metal in the lattice.

CIE 0620 June '04 Paper 3 Q5

3 Zinc is extracted from zinc blende, ZnS.

a Zinc blende is heated in air to give zinc oxide (ZnO) and sulfur dioxide (SO$_2$). Write the equation for this reaction.

b Some of the zinc oxide was mixed with an excess of carbon and heated to 1000 °C. Zinc distils out of the furnace.

$$2ZnO + C \rightleftharpoons 2Zn + CO_2$$
$$C + CO_2 \rightarrow 2CO$$

i Name the two changes of state involved in the process of distillation.

ii Why is it necessary to use an excess of carbon?

iii Name the two chemicals that are **reduced** in these reactions.

c Give two uses of zinc. *CIE 0620 November '07 Paper 3 Q4*

Alternative to practical

1 An investigation was carried out into the reactions of aqueous copper(II) sulfate with magnesium, iron and zinc.

By using a measuring cylinder, 5 cm^3 of aqueous copper(II) sulfate was added to each of three test-tubes. The initial temperature of the solution was measured. Zinc powder was added to the first test-tube, iron powder to the second tube and magnesium powder to the third tube. The mixtures were stirred with the thermometer. All the observations were recorded and the maximum temperature reached was measured. The temperatures are shown on the thermometer drawings below.

Table of results

Metal added	Temperature of solution/°C		Observations
	Initial	Maximum	
zinc	25 (30–20)	58 (60–50)	moderate effervescence, solution paler, brown solid
iron	21 (30–20)	43 (45–35)	little effervescence, brown solid
magnesium	21 (30–20)	73 (75–65)	rapid effervescence, pops with lighted splint, brown solid

a Record the temperature difference for each metal.

b Use your results and observations to answer the following questions.

i Which metal is the most reactive with aqueous copper(II) sulfate?

ii Give two reasons why you chose this metal.

iii Identify the gas given off when magnesium reacts with aqueous copper(II) sulfate. *CIE 0620 November '06 Paper 6 Q6*

17 Water and air

The big picture

- Water, and air, are two key natural resources.
- We use water from rivers and other sources for our water supply. It is cleaned up before being piped to homes.
- 'Clean' air is a natural mixture of gases. We can separate them out, and make use of them.
- We allow harmful waste gases to escape into the air, and these cause a range of problems.
- Oxygen and moisture in the air together cause iron to rust (corrode). Preventing this is a big challenge.

17.1 Water

We all need water ...

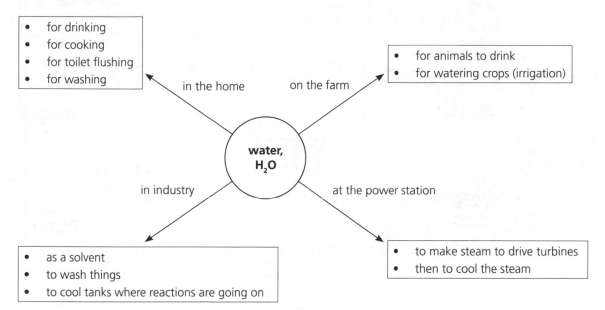

- for drinking
- for cooking
- for toilet flushing
- for washing

in the home

on the farm

- for animals to drink
- for watering crops (irrigation)

water, H₂O

in industry

at the power station

- as a solvent
- to wash things
- to cool tanks where reactions are going on

- to make steam to drive turbines
- then to cool the steam

A chemical test for water

Add a liquid to one of these compounds. If there is a colour change, the liquid contains water:

Compound	anhydrous copper(II) sulfate, $CuSO_4$	anhydrous cobalt(II) chloride, $CoCl_2$
Colour change on adding water	white $\longrightarrow$ blue	blue $\longrightarrow$ pink
Equation	$CuSO_4\ (s)\ +\ 5H_2O\ (l)\ \longrightarrow\ CuSO_4.5H_2O\ (s)$	$CoCl_2\ (s)\ +\ 6H_2O\ (l)\ \longrightarrow\ CoCl_2.6H_2O\ (s)$
Note	*Anhydrous means without water*. Adding water turns the anhydrous compound to the *hydrated* compound. The water molecules are called **water of crystallisation**.	

If the liquid boils at 100°C and freezes at 0°C, then it is pure water.

Our water supply

We depend on rain for our water supply. The rainwater runs over and through the ground, and reaches rivers, reservoirs, and natural underground stores called **aquifers**. We take our water from these.

Making water fit to drink

Before the water is piped to homes, it is treated to make it fit to drink. This involves two main processes:

* removal of solid particles by filtering and skimming. See stages **1 – 5** in the diagram below.
* killing bacteria and other harmful microbes, using chlorine. See **6**.

A modern water treatment plant

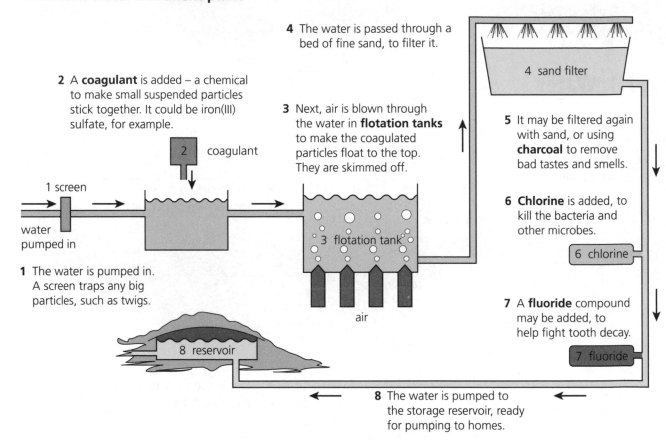

4 The water is passed through a bed of fine sand, to filter it.

4 sand filter

2 A **coagulant** is added – a chemical to make small suspended particles stick together. It could be iron(III) sulfate, for example.

2 coagulant

3 Next, air is blown through the water in **flotation tanks** to make the coagulated particles float to the top. They are skimmed off.

5 It may be filtered again with sand, or using **charcoal** to remove bad tastes and smells.

1 screen

water pumped in

3 flotation tank

6 **Chlorine** is added, to kill the bacteria and other microbes.

6 chlorine

1 The water is pumped in. A screen traps any big particles, such as twigs.

air

7 A **fluoride** compound may be added, to help fight tooth decay.

7 fluoride

8 reservoir

8 The water is pumped to the storage reservoir, ready for pumping to homes.

✓

Quick check for 17.1 (Answers on page 169)

1 Water is used as a *solvent*. What does that mean?
2 Why is water needed in a power station?
3 Describe a test for water.
4 Blue cobalt(II) chloride is an *anhydrous* compound. What does that mean?
5 In the treatment of water, to make it fit to drink:
   a why is filtering needed?   **b** how are bacteria removed?

17.2 Air

What is in it?

Air is a **mixture of gases.** It is mainly nitrogen and oxygen. This shows what is in 'clean' air:

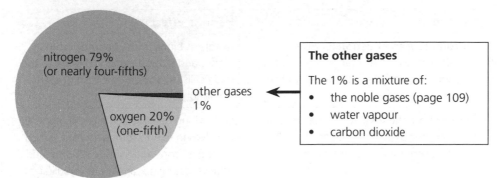

nitrogen 79% (or nearly four-fifths)

oxygen 20% (one-fifth)

other gases 1%

The other gases

The 1% is a mixture of:
- the noble gases (page 109)
- water vapour
- carbon dioxide

Separating the mixture

Extended

The gases can be separated by **fractional distillation** (page 9).
1 First the air is cooled down until it becomes a liquid.
2 Then it is warmed up slowly. This allows the gases to boil off separately, since they all have different boiling points.
3 The gases are put into tanks or cylinders under pressure, and sold for different uses.

The fractional distillation of liquid air

warm up the liquid air

oxygen boils off at	−183 °C
nitrogen boils off at	−196 °C
air is a liquid at	−200 °C

Making use of oxygen

Oxygen is used:
- in oxygen tents in hospitals, for people with breathing difficulties
- along with the hydrocarbon acetylene (ethene) in torches for cutting and welding metal; the burning mixture is so hot that it can melt steel
- as an oxygen supply for astronauts and deep-sea divers
- in steel works, to remove carbon from pig iron (page 119).

Making use of nitrogen

Unlike oxygen, nitrogen is inert (unreactive). It also has a very low boiling point.
These properties lead to many uses. For example it is used:
- inside food packaging, to protect food from oxidation by the oxygen in air
- to rinse out fuel storage tanks, where a mixture of air and fumes could be explosive
- to freeze liquid in pipelines, allowing the pipes to be repaired
- to freeze food, and keep containers of food frozen during transport.
Nitrogen is also used to make ammonia (page 132).

✓

Quick check for 17.2 *(Answers on page 169)*

1 Copy this paragraph, choosing the correct term from each pair in brackets.
Air is a (compound/mixture). The main gas in it is (oxygen/nitrogen), which is (inert/highly reactive). This gas makes up nearly (80%/100%) of the air.
Most of the gases in clean air are (elements/compounds), and all are (metals/non-metals).
2 Air also contains small amounts of the noble gases. Name four noble gases.
3 Nitrogen and oxygen can be separated from the air, by fractional distillation. Explain *why* this is possible.
4 Give three uses of the oxygen obtained by fractional distillation.

17.3 Air pollution

The pie chart in 17.2 showed clean air. But we release many harmful gases into the atmosphere, causing **air pollution**. These are the main pollutants:

Pollutant	Source	Harmful effect
carbon monoxide, CO, a colourless gas with no smell	**incomplete combustion** (burning in insufficient air) of substances that contain carbon; for example petrol in car engines	deadly: it binds to the **haemoglobin** in blood, and stops it carrying oxygen to the body cells, so you can die from oxygen starvation
sulfur dioxide, SO$_2$ acidic gas	mainly from the fossil fuels (coal, oil, and gas) burned in power stations; they all contain some sulfur compounds	causes lung damage and breathing problems; dissolves in rain to form **acid rain**, which damages crops and forests, kills fish, and attacks stonework and metal in buildings
oxides of nitrogen, such as NO, NO$_2$, N$_2$O; called NO$_x$ for short; some are acidic	the nitrogen and oxygen in air react together inside car engines – see below	cause lung damage and breathing problems; some form acid rain
lead compounds	lead is used in many industries; lead compounds were once added to petrol to help it burn smoothly	lead harms the body's nervous system; it can cause brain damage and behavioural problems

Oxides of nitrogen in car exhausts
Air gets so hot inside car engines that the nitrogen and oxygen react together:

nitrogen + oxygen $\xrightarrow{\text{hot car engine}}$ a mixture of nitrogen oxides

Removing the nitrogen oxides
All modern car exhausts contain **catalytic converters**. These contain a catalyst that catalyses the conversion of the oxides back to nitrogen and oxygen again:

nitrogen oxides $\xrightarrow{\text{catalyst in exhaust}}$ nitrogen + oxygen

The diagram on the right shows a typical catalytic converter.

In A: the nitrogen oxides are **reduced**. For example:
$2NO\ (g) \longrightarrow N_2\ (g) + O_2\ (g)$
The products flow to B.

In B: the oxygen from A **oxidises** other harmful gases, for example carbon monoxide:
$2CO\ (g) + O_2\ (g) \longrightarrow 2CO_2\ (g)$
The harmless products then flow out the exhaust.

The catalysts: they are usually transition metals. They are coated onto a ceramic support to give a large surface area for the reactions.

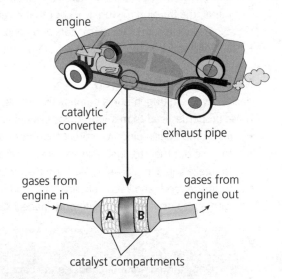

engine

catalytic converter

exhaust pipe

gases from engine in

gases from engine out

A B

catalyst compartments

✔
Quick check for 17.3 *(Answers on page 169)*
1   Carbon monoxide is called a pollutant. Why?
2   What is the source of the carbon monoxide in the atmosphere?
3   What problems are caused by acid rain?
4   **a** What does a catalytic converter do?   **b** Why is it called *catalytic*?
5   Give one example of a reaction that removes a pollutant, in a catalytic converter.

Extended

17.4 Carbon dioxide and the carbon cycle

The carbon cycle

Carbon dioxide continually moves between the atmosphere, living things, and dead remains:

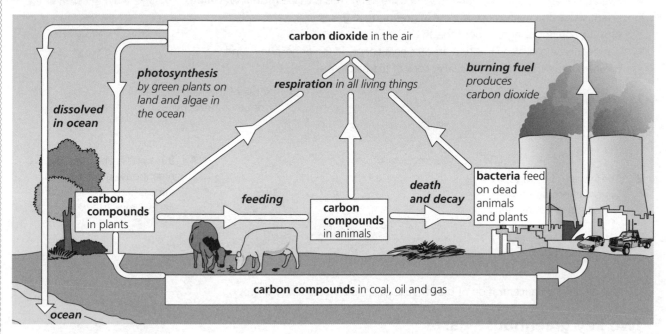

Notes on the processes

Carbon dioxide is added to the atmosphere by ...

1   **Respiration**. This is the process that takes place in the cells of living things, to provide energy. It uses glucose, and produces carbon dioxide, which goes into the air. (We humans breathe it out.)

$C_6H_{12}O_6$ (aq) + 6O_2 (g) $\longrightarrow$ 6CO_2 (g) + 6H_2O (l) + energy
glucose

Bacteria also produce carbon dioxide through respiration, using compounds from dead remains.

2   **Combustion** of fossil fuels. For example when natural gas (methane) burns, the reaction is:

CH_4 (g) + 2O_2 (g) $\longrightarrow$ CO_2 (g) + 2H_2O (l)

Carbon dioxide is removed from the atmosphere by ...

1   **Photosynthesis**. Plants convert carbon dioxide from the air, and water from soil, to glucose:

$$6CO_2\ (g)\ +\ 6H_2O\ (l)\ \xrightarrow[\text{energy from sunlight}]{\text{chlorophyll}}\ C_6H_{12}O_6\ (s)\ +\ 6O_2\ (g)$$
glucose

This process occurs in all green plants, including the tiny algae floating in the ocean. Then, over many millions of years, some dead vegetation turns into coal. Dead algae, and the remains of the organisms that eat them, turn into oil and gas under the ocean.

2   **Dissolving in the ocean**. Some carbon dioxide from the air is absorbed by the ocean. (A balance is reached between the concentration in the air and in the ocean.)

✓

Quick check for 17.4   *(Answers on page 169)*

1   Give two ways in which carbon dioxide:
    a enters the atmosphere   **b** is removed from the atmosphere
2   Describe what happens during: **a** respiration   **b** photosynthesis
3   Which part of the carbon cycle, if any, is *not* natural? Explain your answer.

17.5 Global warming

What is global warming?
- Average air temperatures around the world are rising. This is called **global warming**.
- Scientists say the main cause is a rise in **greenhouse gases** in the atmosphere. (But they do not all agree about this.)
- A greenhouse gas is any gas that absorbs heat energy in the atmosphere, and prevents it from escaping into space. Look at this diagram:

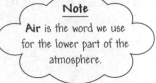

Note
Air is the word we use for the lower part of the atmosphere.

1 The sun sends out energy as light and UV rays.

2 These warm the Earth, which reflects some of the energy again as heat.

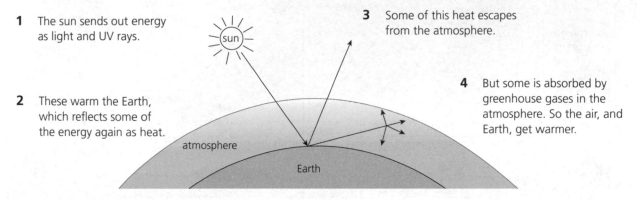

3 Some of this heat escapes from the atmosphere.

4 But some is absorbed by greenhouse gases in the atmosphere. So the air, and Earth, get warmer.

- The higher the concentration of greenhouse gases, the warmer the air grows.

Two key greenhouse gases
This table shows the two main greenhouse gases that are causing concern.

Greenhouse gas	Human activities that increase its level
carbon dioxide, CO$_2$	• combustion of fossil fuels and other carbon-based fuels, in power stations, factories, car engines, and homes • cement production, in which limestone breaks down, giving off carbon dioxide
methane, CH$_4$	• putting decaying organic matter in landfill sites; as the buried material rots, methane forms and can escape to the air • livestock farming; cattle and sheep 'belch' methane from their digestive systems

Predicted consequences
No matter what causes global warming, the consequences are expected to be serious:
- The rising temperatures will cause ice to melt in the polar regions.
- When land ice melts, the water will flow to the sea, causing rising sea levels. So low-lying coastal areas will flood.
- The rising temperatures will cause changes in wind and rainfall patterns too. So climates will change around the world.
- There will be more frequent and severe storms, floods, heatwaves, and drought.
- Climate changes will lead to changes in farming and many other activities.
- Animals and plants that cannot adapt to the changing climates will die out.

✓
Quick check for 17.5   (*Answers on page 169*)
1 What is a *greenhouse gas*? Give two examples.
2 What does the term *global warming* mean?
3 Explain the link between global warming and our use of fossil fuels.
4 Many countries have agreed to reduce their emissions of carbon dioxide, to try to limit global warming. Why are they so concerned about global warming?

17.6 Rusting

What is rusting?
- **Rusting** is the special name for the **corrosion** of **iron**.
- It is an **oxidation**. Oxygen from the air reacts with iron in the presence of moisture.
- The product is hydrated iron(III) oxide, $Fe_2O_3.H_2O$, a brown flaky solid.

Ways to prevent rusting
We use iron for large structures such as bridges and ships, and small ones such as railings and outdoor benches. Rusting weakens iron structures, and can destroy them. So rust prevention is a big challenge. There are three ways to prevent rusting:

1 Stop oxygen (air) from reaching the iron
The iron is coated with something that keeps oxygen out:

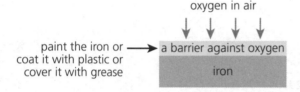

2 Sacrifice another metal in place of iron
A more reactive metal will prevent iron from rusting. For example, magnesium is above iron in the **reactivity series** (page 115). So blocks of magnesium are attached to the sides of ships, and the legs of offshore oil rigs, as shown in the diagram on the right.

Without magnesium: the iron is oxidised. (It loses electrons.) The half equation is –
$$Fe \longrightarrow Fe^{2+} + 2e^-$$
With magnesium: the magnesium is oxidised instead of the iron –
$$Mg \longrightarrow Mg^{2+} + 2e^-$$
So the iron remains unchanged. Magnesium has been sacrificed to protect the iron. This is called **sacrificial protection**.

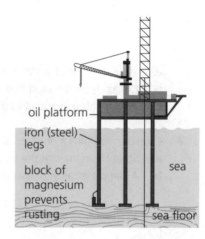

Sacrifical protection for an oil rig. Oil rigs and ships are especially at risk since rusting is faster in salty water.

3 Galvanising
This is a combination of methods 1 and 2 above. The iron is coated with a layer of zinc – which is above iron in the reactivity series – to give **galvanised iron**.

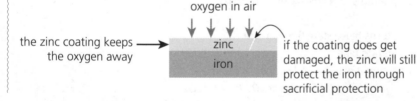

✓

Quick check for 17.6 (*Answers on page 169*)

1 What special name is given to the corrosion of iron?
2 What is the chemical name, and formula, of rust?
3 Why does coating iron with plastic prevent rusting?
4 Iron will not rust in water that has been boiled for five minutes. Why not?
5 Write a half equation to show that rusting is an oxidation reaction.
6 Explain what *sacrificial protection* is.
7 What metal is used in galvanising, and how does galvanising work?

Extended

Questions on section 17

Answers for these questions are on page 169.

Core curriculum

1 The diagram shows a water treatment works.

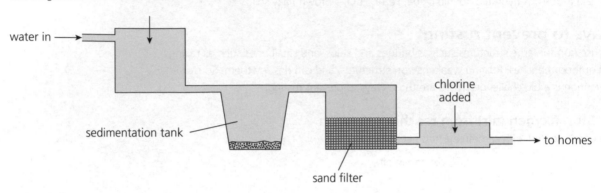

a State one use of water in industry.
b Explain how the sand filter helps purify the water.
c Why is chlorine added to the water?   *CIE 0620 June '08 Paper 2 Q4*

2 Clean air contains a number of different gases.
   a State the names of the two gases which make up most of the air.
   b A sample of air is drawn through the apparatus shown below.

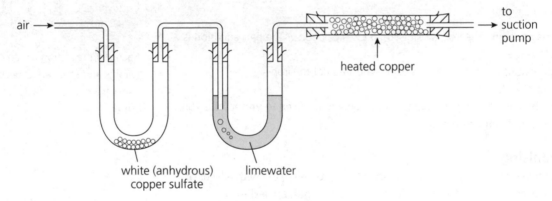

   i When the air is drawn through the apparatus, the limewater turns milky. Which gas turns limewater milky?
   ii The white (anhydrous) copper sulfate turns blue. State the name of the substance which turns white copper sulfate blue.
   iii Oxygen is removed from the air by passing it over heated copper. Write the equation for this reaction.   *CIE 0620 June '07 Paper 2 Q2*

3 Two elements, the metal lead and the non-metal sulfur, are both found as pollutants in air.
   a **i** What is the source of the small amount of lead present in the air?
   ii State an adverse effect of lead on health.
   b Explain why burning fossil fuels containing sulfur is harmful to the environment.
           CIE 0620 June '07 Paper 2 Q1

Extended curriculum

1 Titanium is very resistant to corrosion. One of its uses is as an electrode in the cathodic protection of large steel structures from rusting.

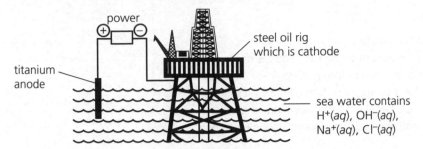

a Define oxidation in terms of electron transfer.

b The steel oil rig is the cathode. Name the gas formed at this electrode.

c Name the **two** gases formed at the titanium anode.

d Explain why the oil rig does not rust.

e Another way of protecting steel from corrosion is sacrificial protection.
Give **two** differences between sacrificial protection and cathodic protection.

CIE 0620 June '07 Paper 3 Q5

Alternative to practical

1 The diagram shows the rusting of a sample of iron filings.

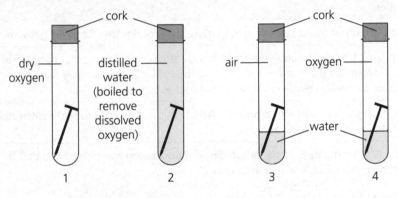

a Predict the order in which rust would appear:

   i first   **ii** second

b Explain your prediction.   *CIE 0620 June '04 Paper 6 Q2*

2 The diagram shows the rusting of a sample of iron filings.

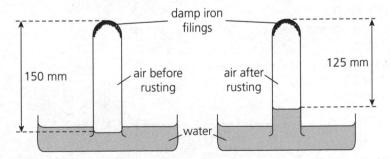

a Describe the appearance of the iron after rusting.

b **i** Why does the water rise up the tube?

   ii Calculate the percentage change in the volume of air in the tube.

c What difference would be observed if

   i an iron nail was suspended in the tube instead of using iron filings,

   ii the water contained salt?

CIE 0620 November '07 Paper 6 Q2

18 Some non-metal elements and compounds

The big picture

- Nitrogen, sulfur and carbon are all non-metals.
- Some of their compounds are very important to industry, and the economy.
- Nitrogen is used to make ammonia, which is in turn used to make nitric acid and fertilisers. Fertilisers are essential for growing crops to feed the world.
- Sulfur is used to make sulfuric acid, a key chemical in many industries.
- The manufacture of sulfuric acid and ammonia have something in common: they involve reversible reactions.
- Carbon occurs naturally as carbonates, in rock such as limestone. Limestone has many important uses, from road building to controlling acidity in soil.

18.1 Nitrogen, ammonia, and fertilisers

Four big questions about nitrogen

Question	Answer
1 Why is nitrogen so important?	Crops and other plants use nitrogen to make the proteins they need for **growth**.
2 Where do plants get their nitrogen from?	Nitrogen is taken in through the roots, mostly as nitrates. These occur naturally in soil. But they get used up by crops, leaving the soil worn out. So they must be replaced by nitrates made in factories, and sold as fertilisers.
3 How is nitrogen turned into nitrates, to make fertilisers?	First, nitrogen is turned into **ammonia (NH$_3$)**. This is then used to make **nitric acid**. **Nitrates** are salts of nitric acid.
4 How is ammonia made?	In industry, ammonia is made by combining nitrogen and hydrogen. Note that the reaction is reversible, and a calatyst is used: $$N_2\ (g)\ +\ 3H_2\ (g)\ \underset{\xleftarrow{\hspace{2cm}}}{\overset{\text{iron as catalyst}}{\xrightarrow{\hspace{2cm}}}}\ 2NH_3\ (g)$$

The Haber process for making ammonia

The method used to make ammonia in industry is called the **Haber process**.

The raw materials: nitrogen and hydrogen

Nitrogen from ...
- the air. The oxygen is removed by burning hydrogen in air, leaving nitrogen behind: $$\underbrace{N_2\ (g)\ +\ O_2\ (g)}_{\text{air}}\ +\ 2H_2\ (g)\ \longrightarrow\ \underset{\text{unchanged}}{N_2\ (g)}\ +\ H_2O\ (l)$$

Hydrogen from ...
- the reaction between methane (natural gas) and steam: $$CH_4\ (g)\ +\ 2H_2O\ (g)\ \longrightarrow\ CO_2\ (g)\ +\ 4H_2\ (g)$$ - or the cracking of hydrocarbons such as ethane (page 146): $$C_2H_6\ (g)\ \longrightarrow\ C_2H_4\ (g)\ +\ H_2\ (g)$$ Both reactions require catalysts.

The reaction conditions

Look again at the reversible reaction for making ammonia:

$$N_2 (g) + 3H_2 (g) \underset{endothermic}{\overset{exothermic}{\rightleftharpoons}} 2NH_3 (g)$$

4 molecules   2 molecules

In a closed system the reaction reaches equilibrium, with a mixture of ammonia, nitrogen, and hydrogen present. The aim is shift the equilibrium to the right, giving more ammonia. Conditions are chosen to favour the forward reaction *and* make it fast enough.

These are the optimum conditions:

Condition		Comment
pressure	high (200 atmospheres)	High pressure favours the side of the equation with fewer molecules – the ammonia.
temperature	moderate (450° C) - a compromise!	• The forward reaction is exothermic, so a low temperature favours it. • But at low temperatures, the reaction is too slow.
catalyst	iron	Speeds up the forward and backward reactions equally. So equilibrium is reached faster. (But a catalyst does not shift equilibrium.)
remove product	the reaction mixture is cooled to remove ammonia as a liquid	Removing the ammonia prevents it from breaking down to nitrogen and hydrogen again.
recycle	unreacted gases are recycled	The gases are given another chance to react at the catalyst, so the overall yield improves.

Making ammonia in the lab

Ammonia forms when **ammonium salts** are heated with **sodium hydroxide**. For example:

ammonium chloride + sodium hydroxide $\longrightarrow$ sodium chloride + water + ammonia

$$NH_4Cl (s) + NaOH (s) \longrightarrow NaCl (s) + H_2O (l) + NH_3 (g)$$

This is the displacement of a weak base (ammonia) by a strong base (sodium hydroxide).
- Ammonia is an alkaline gas, so it will turn damp red litmus paper blue.
- It is very soluble in water, giving the alkali ammonium hydroxide.

More about fertilisers

- Plants need a good supply of **nitrogen**, **potassium**, and **phosphorus** from the soil.
- They use these up fast, so fertilisers must be added to the soil to replace them.
- Most fertilisers are salts, made by neutralising an acid with an alkali. Look at this table:

Plant nutrient	What it does	A fertiliser to provide it	Made by neutralising ...
nitrogen, N	see earlier	ammonium nitrate, NH_4NO_3 ammonium sulfate, $(NH_4)_2SO_4$	nitric acid with ammonia solution sulfuric acid with ammonia solution
potassium, K	among other things, it helps to protect plants against disease	potassium nitrate, KNO_3	(occurs naturally as **saltpetre**)
phosphorus, P	it improves crop yields; it helps roots to grow and crops to ripen	ammonium phosphate, $(NH_4)_3PO_4$	phosphoric acid with ammonia solution

✓
Quick check for 18.1   *(Answers on page 169)*

1   Give the sources of the raw materials for the Haber process.
2   Explain why high pressure is used, in the Haber process.
3   Which two steps are taken to ensure a decent *reaction rate* in the Haber process?
4   How is ammonia prepared in the laboratory?
5   How does potassium nitrate fertiliser help plants?

18.2 Sulfur and sulfuric acid

Extended

Sulfur

- The element **sulfur** is a non-metal. Its symbol is S.
- It is quite a common element in the Earth's crust.
- It is a **yellow solid** at room temperature.
- It has a **molecular** structure. Each molecule is made up of **8** sulfur atoms (S_8).
- Since it is molecular, it has a quite a **low melting point** (115°C)

Sources and uses

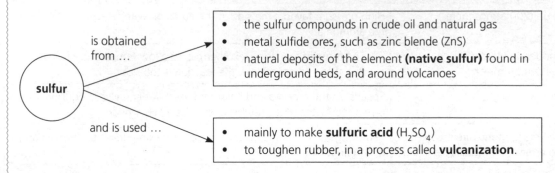

sulfur

is obtained from …

- the sulfur compounds in crude oil and natural gas
- metal sulfide ores, such as zinc blende (ZnS)
- natural deposits of the element **(native sulfur)** found in underground beds, and around volcanoes

and is used …

- mainly to make **sulfuric acid** (H_2SO_4)
- to toughen rubber, in a process called **vulcanization**.

Sulfuric acid (H_2SO_4)

Sulfuric acid is a very important industrial chemical. It is made in enormous quantities by the **Contact Process**. The raw materials are sulfur and air.

The Contact Process

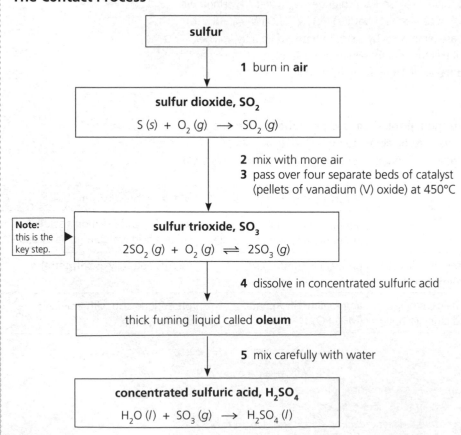

sulfur

1 burn in **air**

sulfur dioxide, SO_2

$$S\ (s) + O_2\ (g) \longrightarrow SO_2\ (g)$$

2 mix with more air
3 pass over four separate beds of catalyst (pellets of vanadium (V) oxide) at 450°C

Note: this is the key step.

sulfur trioxide, SO_3

$$2SO_2\ (g) + O_2\ (g) \rightleftharpoons 2SO_3\ (g)$$

4 dissolve in concentrated sulfuric acid

thick fuming liquid called **oleum**

5 mix carefully with water

concentrated sulfuric acid, H_2SO_4

$$H_2O\ (l) + SO_3\ (g) \longrightarrow H_2SO_4\ (l)$$

Look at the key step: it is a reversible reaction. So the reaction mixture contains sulfur dioxide and oxygen as well as sulfur trioxide. Reaction conditions must be chosen to shift the equilibrium to the right, giving as much sulfur trioxide as possible, for a reasonable cost.

Extended

The reaction conditions

Here is the key reaction again:

$$2SO_2\ (g)\ +\ O_2\ (g)\ \underset{\text{endothermic}}{\overset{\text{exothermic}}{\rightleftharpoons}}\ 2SO_3\ (g)$$

3 molecules   2 molecules

Make the link to ...
shifting equilibrium,
on page 75.

These are the optimum conditions to give an acceptable yield of sulfur trioxide, fast enough, and at a reasonable cost:

Condition	Effect on the forward reaction of changing this condition	Final condition chosen
pressure	In the forward reaction 3 molecules $\rightarrow$ 2 molecules so **raising** the pressure will favour it.	**normal pressure** Gives an acceptable yield without going to the expense of fitting a high-pressure system.
temperature	The forward reaction is exothermic, so **lowering** the temperature will favour it.	**moderate temperature (450°C)** This is a compromise: at a lower temperature, the reaction is too slow.
catalyst	A catalyst speeds up the forward and backward reactions equally. So equilibrium is reached faster.	a catalyst of **vanadium (V) oxide** This is expensive, but worth the cost.

Uses of sulfuric acid

Sulfuric acid is used in car batteries. It is also a really important chemical in industry:

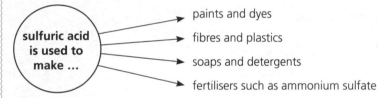

sulfuric acid is used to make ...
→ paints and dyes
→ fibres and plastics
→ soaps and detergents
→ fertilisers such as ammonium sulfate

Chemical reactions of dilute sulfuric acid

Dilute sulfuric acid shows typical acidic reactions:

- it turns universal indicator red
- it reacts with **metals** to form **salts** (sulfates) and **hydrogen**
- it reacts with **metal oxides and hydroxides (bases)** to form **salts** (sulfates) and **water**
- it reacts with **metal carbonates** to form **salts** (sulfates), **water** and **carbon dioxide**

Sulfur dioxide

Sulfur dioxide, SO_2, is a colourless gas. It is manufactured by burning sulfur in air.
It has several important uses:

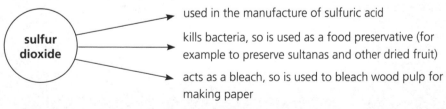

sulfur dioxide
→ used in the manufacture of sulfuric acid
→ kills bacteria, so is used as a food preservative (for example to preserve sultanas and other dried fruit)
→ acts as a bleach, so is used to bleach wood pulp for making paper

Sulfur dioxide also forms when fossil fuels burn, since these contain sulfur as an impurity. It is converted to sulfuric acid in damp air, so is a key cause of **acid rain** (page 126).

✓
Quick check for 18.2   (Answers on page 169)
1   Name two sources of sulfur.
2   Give the essential conditions for converting sulfur dioxide to sulfur trioxide, in the Contact Process.
3   Why is sulfur dioxide used in: **a** the paper industry?   **b** the food industry?

18.3 Calcium carbonate

What are carbonates?

- Carbonates are compounds that contain the carbonate ion, CO_3^{2-}.
- The rocks **limestone**, **chalk**, and **marble** are all mainly calcium carbonate ($CaCO_3$).
- Limestone and chalk were formed over millions of years, under the ocean, from shells and skeletons of sea creatures. High pressure converted some limestone to marble.

> **Test for carbonates in rock**
> - Drip acid onto a sample of the rock.
> - If it fizzes, the rock contains a carbonate. (The fizz is due to carbon dioxide gas.)

Uses of limestone

Limestone is mined at quarries. It has many uses:

Where it is used	What for?
in the steel industry	to remove acidic impurities when extracting iron and making steel (pages 118 – 119)
in the building industry	to make **cement**; limestone is heated with clay, then gypsum (calcium sulfate) is added
on the farm	to neutralise acidity in soil (page 86)
at the power station	to remove acidic flue gases such as sulfur dioxide, produced when the sulfur compounds in fossil fuels burn; the removal of sulfur dioxide from flue gases is called **desulfurisation**

Quicklime and slaked lime

Limestone is also a **raw material** for two other important chemicals:

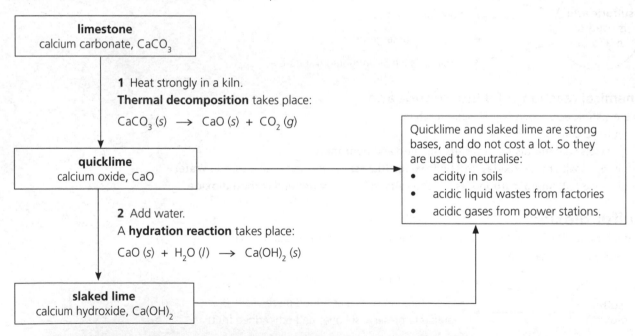

limestone
calcium carbonate, $CaCO_3$

1 Heat strongly in a kiln.
Thermal decomposition takes place:

$$CaCO_3\ (s)\ \longrightarrow\ CaO\ (s)\ +\ CO_2\ (g)$$

quicklime
calcium oxide, CaO

2 Add water.
A **hydration reaction** takes place:

$$CaO\ (s)\ +\ H_2O\ (l)\ \longrightarrow\ Ca(OH)_2\ (s)$$

slaked lime
calcium hydroxide, $Ca(OH)_2$

Quicklime and slaked lime are strong bases, and do not cost a lot. So they are used to neutralise:
- acidity in soils
- acidic liquid wastes from factories
- acidic gases from power stations.

✓
Quick check for 18.3 *(Answers on page 169)*

1. What is the formula of the carbonate ion?
2. Name three rocks that contain calcium carbonate.
3. Name three uses of limestone which rely on its ability to react with acids.
4. How is cement manufactured from limestone?
5. How does *slaked lime* differ chemically from *quicklime*?
6. Calcium hydroxide is often chosen *for desulfurisation*.
   a What does the term in italics mean?   **b** Why is calcium hydroxide suitable?

Questions on section 18

Answers for these questions are on page 169.

Core curriculum

1 Fertilisers often contain ammonium nitrate.

 a What effect do fertilisers have on crops?

 b Name one metal ion which is commonly present in fertilisers.

 c Which one of the following ions is commonly present in fertilisers?

 bromide, *chloride*, *hydroxide*, *phosphate*

 d Ammonium nitrate can be made by adding nitric acid to a solution of ammonia. What type of reaction is this?

 e Which two of the following statements about ammonia are true?

 i ammonia is insoluble in water

 ii ammonia turns red litmus blue

 iii a solution of ammonia in water has a pH of 7

 iv ammonia has a molecular structure   *CIE 0620 November '04 Paper 2 Q5*

2 The diagram shows a statue in a park in an industrial town. The statue is made from limestone.

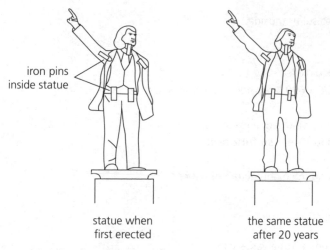

| statue when first erected | the same statue after 20 years |

 iron pins inside statue

 a State the name of the chemical present in limestone.

 b Use ideas about the chemistry of atmospheric pollutants to suggest how and why the statue changes over 20 years.

 c Parts of the statue are joined together with iron pins. After 30 years, the arm falls off the statue. Suggest why the arm falls off.   *CIE 0620 June '08 Paper 2 Q2*

Extended curriculum

1 Ammonia contains the elements nitrogen and hydrogen. It is manufactured from these elements in the Haber process.

$$N_2 (g) + 3H_2 (g) \rightleftharpoons 2NH_3 (g)$$

The forward reaction is exothermic.

 a Name two raw materials from which hydrogen is manufactured.

 b The table shows how the percentage of ammonia in the equilibrium mixture varies with pressure, at 600°C.

Percentage ammonia	8	12	15	20
Pressure/atm	200	300	400	500

 Explain why the percentage of ammonia increases as the pressure increases.

Extended

  c **i** What is the catalyst for this reaction?

      ii Newer catalysts have been discovered for this process. Using these catalysts, the operating temperature is lowered from 450°C to 400°C. What is the advantage of using a lower temperature? Explain your answer.

  d After passing over the catalyst, the mixture contains 15% of ammonia. It is cooled and the ammonia liquefies and is separated from the unreacted nitrogen and hydrogen. They are recycled.

      i How are the gases recycled?

      ii Only the ammonia gas liquefies. Suggest an explanation for this.

  e Urea, $CO(NH_2)_2$, is one of the fertilisers manufactured from ammonia. Ammonia is heated with carbon dioxide.

      i Write an equation for the manufacture of urea.

      ii Explain why urea on its own might not be very effective in promoting crop growth.
                            CIE 0620 November '03 Paper 3 Q1, Q5

                                        CIE 0620 November '06 Paper 3 Q5

2 Sulfur dioxide is a by-product of the extraction of zinc from zinc blende (zinc sulfide). Most of the sulfur dioxide is used to make sulfur trioxide. This is used to manufacture sulfuric acid. Some of the acid is used in the plant, but most of it is used to make fertilisers.

  a Describe the appearance of sulfur dioxide.

  b Give another use of sulfur dioxide.

  c **i** Describe how sulfur dioxide is converted into sulfur trioxide.

      ii Write the equation for the reaction.

  d Name a fertiliser made from sulfuric acid.

  e What environmental problem is caused by sulfur dioxide?
                                  CIE 0620 November '07 Paper 3 Q4

3 Sulfuric acid is made by the Contact process in the following sequence of reactions.

sulfur $\rightarrow$ **sulfur dioxide** $\rightarrow$ **sulfur trioxide** $\rightarrow$ **sulfuric acid**

  a **i** How is sulfur dioxide made from sulfur?

      ii Sulfur dioxide has other uses. Why is it used in the manufacture of paper?

      iii How does it preserve food?

  b The equation for a stage of the Contact process is:

      $2SO_2\,(g) + O_2\,(g) \rightleftharpoons 2SO_3\,(g)$

      The percentage of sulfur trioxide in the equilibrium mixture varies with temperature.

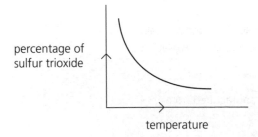

      i Does the percentage of sulfur trioxide in the equilibrium mixture *increase, stay the same* or *decrease* as the temperature increases?

      ii Is the forward reaction in the equilibrium $2SO_2 + O_2 \rightleftharpoons 2SO_3$ exothermic or endothermic? Give a reason for your choice.

      iii Explain, mentioning both rate and percentage yield, why the temperature used in the Contact process is 450°C.

      iv Describe how the sulfur trioxide is changed into concentrated sulfuric acid.
                                  CIE 0620 June '06 Paper 3 Q5

19 Organic compounds, and fuels

The big picture

- Organic chemistry is the study of organic compounds: carbon compounds that originate in living things.
- Carbon – carbon bonds are the backbone of these compounds.
- All the fuels we burn are organic compounds.
- Petroleum is a mixture of hundreds of compounds. It becomes more useful when refined into groups of compounds close in chain length.

19.1 What are organic compounds?

- **Organic compounds** are carbon compounds that originate in **living things**.
- Their key feature is the covalent bonds between carbon atoms, in their molecules. Some organic compounds have thousands of carbon atoms linked in long chains.
- Organic compounds that contain *only* carbon and hydrogen are called **hydrocarbons**.

Some examples of organic compounds

This is a molecule of **methane**: the simplest organic compound of all.

This is **butane**. Like methane, butane is a hydrocarbon.

This is **ethanol**. It contains oxygen so is not a hydrocarbon.

✓
Quick check for 19.1   (Answers on page 169)
1   Is iron(II) oxide an organic compound? Explain your answer.
2   The formula for ethanol is C_2H_5OH. Write formulae for methane and butane.
3   Is it a hydrocarbon?   **a** $C_{20}H_{42}$ **b** $C_6H_{13}NO_2$ **c** $C_{18}H_{34}O_2$

19.2 The fossil fuels

The **fossil fuels** are the remains of living things from millions of years ago. So they are mainly organic compounds:

Fossil fuel	What's in it?	Where it originated
coal	a large mixture of organic compounds	from vegetation buried in ancient swamps
petroleum (oil)	a large mixture of organic compounds	from the remains of sea organisms, buried in the ocean floor
natural gas	mainly methane, plus some butane and other hydrocarbon gases	also from dead sea organisms (it is usually found with petroleum)

- The fossil fuels also contain some sulfur compounds as impurities.
- Coal and natural gas are used directly as fuels. For example both are burned in power stations. Natural gas is used for cooking and heating, in homes.
- But petroleum goes through a refining process before it is used.

Refining petroleum

Petroleum is the most useful fossil fuel. It contains many liquid compounds, and some gases, that can be used as fuels, and as the starting point for plastics and other materials.

- The first step is to **separate** the petroleum into groups of compounds with molecules of similar size. This is called **refining** the petroleum.
- It is carried out by **fractional distillation**.
- The groups of compounds are called **fractions**.
- The separation is carried out in a **fractionating tower**, as in in the drawing below.

> **Remember**
> Fractional distillation works because compounds have different boiling points (See page 9.)

The fractions and their uses

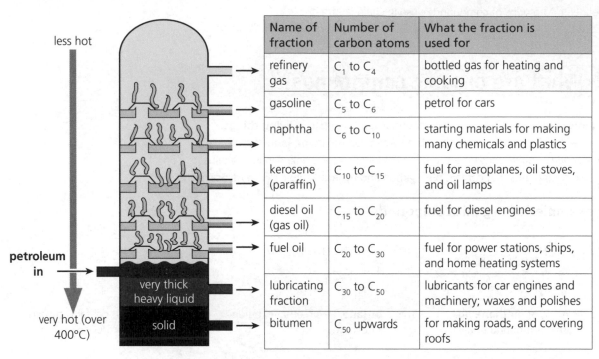

less hot

petroleum in

very hot (over 400°C)

Name of fraction	Number of carbon atoms	What the fraction is used for
refinery gas	C_1 to C_4	bottled gas for heating and cooking
gasoline	C_5 to C_6	petrol for cars
naphtha	C_6 to C_{10}	starting materials for making many chemicals and plastics
kerosene (paraffin)	C_{10} to C_{15}	fuel for aeroplanes, oil stoves, and oil lamps
diesel oil (gas oil)	C_{15} to C_{20}	fuel for diesel engines
fuel oil	C_{20} to C_{30}	fuel for power stations, ships, and home heating systems
lubricating fraction	C_{30} to C_{50}	lubricants for car engines and machinery; waxes and polishes
bitumen	C_{50} upwards	for making roads, and covering roofs

boiling points increase

- Note that the compounds with the smallest molecules have the lowest boiling points and are the most **volatile** – they evaporate the most easily. They boil off first.
- The compounds with the largest molecules remain behind as solids.
- The larger its molecules, the less easily a compound burns. So the lowest two fractions are not used as fuels.

The problems with fossil fuels

- When fossil fuels burn in a good supply of air, they form carbon dioxide and water vapour. For example methane burns like this:

methane + oxygen $\longrightarrow$ carbon dioxide + water + energy

CH_4 (g) + $2O_2$ (g) $\longrightarrow$ CO_2 (g) + $2H_2O$ (l) + energy

But carbon dioxide is a **greenhouse gas** – it helps to trap heat in the atmosphere. Many scientists believe it is the main cause of global warming, and that we have brought on global warming by burning fossil fuels. (But not all scientists agree.)
- The burning of fossil fuels also produces sulfur dioxide and other pollutants.

> **Make the link to ...**
> global warming, on page 128.

> **Make the link to ...**
> air pollution, on page 126.

✓

Quick check for 19.2   *(Answers on page 169)*

1   Petroleum is *refined*. What does that mean, and how is it carried out?
2   One compound in petroleum has 15 carbon atoms. Another has 5. Which has the higher boiling point?
3   Which fraction from refining petroleum is used in road-building?
4   What is the link between fossil fuels and global warming?

Questions on section 19

Answers for these questions are on page 169.

Core curriculum

1 Petroleum is separated into useful fractions by distillation.

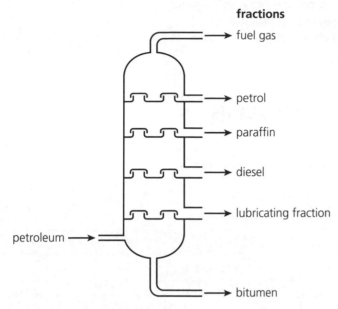

fractions

fuel gas

petrol

paraffin

diesel

lubricating fraction

petroleum →

bitumen

a **i** What do you understand by the term fraction?
  ii Which fraction has the lowest boiling point?
  iii Describe how distillation is used to separate these fractions.
b State a use for
  i the paraffin fraction
  ii the bitumen fraction.

2 A major source of energy is the combustion of fossil fuels.
a **i** Name a solid fossil fuel.
  ii Name a gaseous fossil fuel.
b Petroleum is separated into more useful fractions by fractional distillation.
  i Name two liquid fuels obtained from petroleum.
  ii Name two useful products obtained from petroleum, that are not used as fuels.
  iii Give another mixture of liquids that is separated on an industrial scale by fractional distillation. *CIE 0620 June '07 Paper 3 Q1*

3 Coal, natural gas and petroleum are obtained from the ground for use as fuels.
a What name is used for this group of fuels?
b What is the main compound in natural gas?
c Petroleum is a mixture of *hydrocarbons*.
  i What is a *hydrocarbon*?
  ii How is the mixture separated?
d When the mixture is separated, different fractions are obtained.
Match up the fractions named on the right with their uses.

naphtha	fuel for cars
fuel oil	bottled gas
gasoline	making chemicals
refinery gas	fuel for home heating
kerosene	aircraft fuel

20 Families of organic compounds

The big picture

- There are many thousands of organic compounds. But they fall into families of compounds with similar structures and chemical properties.
- The compounds in a family share chemical properties because they have the same functional group: a small group of atoms that largely dictates their reactions.
- So if you know the reactions of one member of the family, you can predict the reactions of others.

20.1 Some groups of organic compounds

Organic compounds fall into families of compounds with similar structures and properties. Below are four families, and the names and structures of some of their simplest members.

Alkanes

Alkanes are **hydrocarbons** where each carbon forms **4 single bonds**.`

Examples				
Name and formula	methane, CH_4	ethane, C_2H_6	propane, C_3H_8	butane, C_4H_{10}
Molecular structure	H │ H─C─H │ H	H H │ │ H─C─C─H │ │ H H	H H H │ │ │ H─C─C─C─H │ │ │ H H H	H H H H │ │ │ │ H─C─C─C─C─H │ │ │ │ H H H H

If the name of an organic compound ends in –**ane** it is an **alkane**.

Alkenes

Alkenes are **hydrocarbons** that contain a **double bond** between carbon atoms.

Examples			
Name and formula	ethene, C_2H_4	propene, C_3H_6	but-1-ene, C_4H_8
Molecular structure	H H │ │ C═C │ │ H H	H H │ │ C═C─C─H │ │ │ H H H	H H H │ │ │ C═C─C─C─H │ │ │ │ H H H H

Each is named like the alkane with the same number of carbon atoms, but with –**ene** at the end.

Alcohols

Alcohols are like alkanes, but have an **OH group** attached to a carbon atom.

Examples			
Name and formula	ethanol, C_2H_5OH	propan-1-ol, C_3H_7OH	butan-1-ol, C_4H_9OH
Molecular structure	H H │ │ H─C─C─OH │ │ H H	H H H │ │ │ H─C─C─C─OH │ │ │ H H H	H H H H │ │ │ │ H─C─C─C─C─OH │ │ │ │ H H H H

Each is named like the alkane with the same number of carbon atoms, but with –**ol** at the end.

Carboxylic acids

Carboxylic acids contain the **COOH** group: a carbon atom with an **=O** and **−OH** attached to it.

Examples			
Name and formula	ethanoic acid, CH_3COOH	propanoic acid, C_2H_5COOH	butanoic acid, C_3H_7COOH
Molecular structure			

Each is named like the alkane with the same number of carbon atoms, but with **−oic acid** at the end.

✓
> **Quick check for 20.1**   *(Answers start on page 169)*
> 1   Explain how the structure of an alkene differs from that of an alkane.
> 2   An organic compound with 3 carbon atoms contains an OH group.
>   Name a compound it could be.
> 3   Which type of organic compound is: **a** pentan-1-ol?   **b** hexanoic acid?
> 4   Draw the structure of the carboxylic acid with 3 carbon atoms.

20.2 Homologous series

A family of organic compounds is also called a **homologous series**. The compounds in a series have similarities – and differences:

Their similarities	Their differences
• the same **general formula** • the same **functional group** • similar **chemical properties**, because of this functional group	• different **chain lengths** – the chain length increases by 1 carbon atom at a time • different **physical properties**, due to increasing molecular size

Look at the patterns for these four families:

Homologous series	General formula (n = no of carbon atoms)	Functional group	Chemical properties
alkanes	C_nH_{2n+2}	C–C	all are generally unreactive
alkenes	C_nH_{2n}	C=C	all react with bromine
alcohols	$C_nH_{2n+1}OH$	O–H	all react with sodium
carboxylic acids	$C_nH_{2n}O_2$	COOH	all are acidic, and neutralised by sodium hydroxide

Now look at the differences within one family:

Alcohols	No of carbon atoms (n)	Molecular formula	Boiling point °C	
methanol	1	CH_3OH	65	
ethanol	2	C_2H_5OH	79	boiling points increase
propan-1-ol	3	C_3H_7OH	98	
butan-1-ol	4	C_4H_9OH	117	

✓
Quick check for 20.2 *(Answers on page 170)*

1 Members of the same homologous series have similar chemical properties. Why?
2 How does chain length differ, for members of a homologous series?
3 Give the formula for:
   a an alkane with 8 carbon atoms **b** an alkene with 10 carbon atoms
   c an alcohol with 7 carbon atoms **d** a carboxylic acid with 6 carbon atoms
4 Which family does this belong to? **a** C_8H_{16} **b** $C_5H_{11}OH$ **c** $C_{30}H_{62}$
5 See if you can estimate the boiling point for a straight-chained alcohol with 5 carbon atoms.

20.3 Structural isomers

Structural isomers are compounds with the same formula, but a different arrangement of atoms. The longer the carbon chain, the more structural isomers there will be.

Examples from the alkane family

For example there are two alkanes with 4 carbons atoms:

Molecular formula	Type of carbon chain	Molecular structure	Name	Boiling pt (° C)
C_4H_{10}	straight		butane	−0.5
C_4H_{10}	branched		methyl propane	−12

So butane and methyl propane are **structural isomers** (or just **isomers**).

Examples from the alcohol family

For example there are two alcohols with 3 carbon atoms:

Molecular formula	OH attached to ...	Molecular structure	Name	Boiling pt (° C)
C_3H_7OH	end of chain		propan-1-ol	98
C_3H_7OH	middle of chain		propan-2-ol	82

Boiling points of isomers

Look at the boiling points, in the tables above.
- The compounds with branched chains have lower boiling points.
- That is because the branches prevent the molecules from getting close together. So the attractive forces between them are weaker. So less energy is needed to form a gas.

Quick check for 20.3 *(Answers on page 170)*

1. What are *structural isomers*?
2. Alkenes can have their double bonds in different positions, giving isomers. See if you can draw *three* isomers with the formula C_4H_8. (One has a branch.)
3. Find methyl propane, on page 144. Why has it been given that name?

20.4 The alkanes

The alkanes are the simplest family of organic compounds. Their simplest member is methane, CH_4.

- The alkanes are hydrocarbons.
- They form the homologous series with the general formula C_nH_{2n+2}.
- All their carbon atoms form 4 single covalent bonds.

Since all their **carbon–carbon bond**s are **single bonds**, the alkanes are called **saturated**. Saturated compounds are generally **unreactive**. But the alkanes do have some reactions:

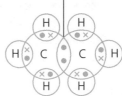

one shared pair of electrons (a single bond)

The bonding in ethane. All the bonds are single bonds.

1 They burn in air.	
Type of reaction	**combustion**
Reactants	alkane + plenty of oxygen (from air) for complete combustion
What else is required	flame to ignite the fuel
What is produced	carbon dioxide + water + plenty of heat
Equation for methane	$CH_4\ (g)\ +\ 2O_2\ (g)\ \longrightarrow\ CO_2\ (g)\ +\ 2H_2O\ (l)$

Note
Incomplete combustion of an alkane gives poisonous **carbon monoxide**, and water.

Extended

2 They react with chlorine.	
Type of reaction	**substitution** (one or more hydrogen atoms are replaced by chlorine atoms)
Reactants	alkane + chlorine gas
What else is required	**UV light** – this is a photochemical reaction
What is produced	hydrogen chloride + chloroalkanes
Equations for methane	$CH_4\ (g)\ +\ Cl_2\ (g)\ \longrightarrow\ CH_3Cl\ (l)\ +\ HCl\ (g)$ then further substitutions can take place: $CH_3Cl\ (l)\ +\ Cl_2(g)\ \longrightarrow\ CH_2Cl_2\ (l)\ +\ HCl\ (g)$ $CH_2Cl_2\ (l)\ +\ Cl_2(g)\ \longrightarrow\ CHCl_3\ (l)\ +\ HCl\ (g)$ $CHCl_3\ (l)\ +\ Cl_2(g)\ \longrightarrow\ CCl_4\ (l)\ +\ HCl\ (g)$

Note their names
CH_3Cl chloromethane
CH_2Cl_2 dichloromethane
$CHCl_3$ trichloromethane
CCl_4 tetrachloromethane

Quick check for 20.4 *(Answers on page 170)*

1. Why are alkanes described as *saturated*?
2. Are alkanes generally *reactive*, or *unreactive*?
3. Write the equation for the combustion of ethane (C_2H_6).
4. Why is the reaction between an alkane and chlorine described as a *substitution*?
5. Another halogen, bromine, reacts in a similar way to chlorine. It is a liquid. Write the equation for the initial reaction between methane and bromine.
6. What happens when bromine is added to hexane in the dark?

20.5 Alkenes

The carbon atoms share two pairs of electrons

The alkenes are a family of organic compounds. The simplest alkene is ethene, C_2H_4:
- The alkenes are hydrocarbons.
- They form the homologous series with the general formula C_nH_{2n}.
- They have a **double covalent bond** between two carbon atoms.

Since their molecules contain **a C=C double bond**, alkenes are called **unsaturated**. This double bond means they are generally **reactive**.

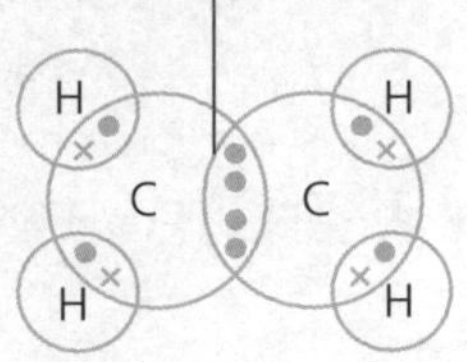

The bonding in ethene.

The bonding in ethene. Note the C=C double bond.

How are alkenes obtained?
- Alkenes are obtained from **alkanes** by a process called **cracking**.
- Cracking involves heating vaporized alkanes in the presence of a **catalyst**.
- Cracking breaks down molecules into smaller ones.
- Cracking always produces some molecules with double bonds: alkenes.

Example 1 The cracking of decane gives an alkane and two alkenes:

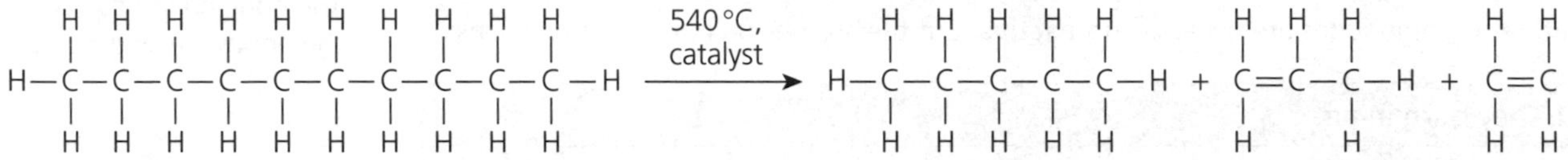

decane, $C_{10}H_{22}$ (from naphtha fraction) → pentane, C_5H_{12} suitable for petrol + propene, C_3H_6 + ethene, C_2H_4 (540 °C, catalyst)

Example 2 The cracking of ethane gives ethene and hydrogen:

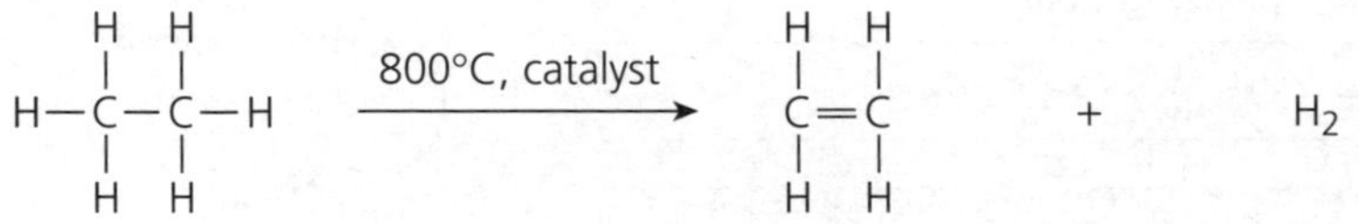

ethane, C_2H_6 → (800°C, catalyst) ethene, C_2H_4 + hydrogen H_2

The ethane for this reaction is obtained from the refinery gas fraction, in refining petroleum (page 140). The hydrogen that is produced could be used to make ammonia (page 132).

Note Ethene is often drawn like this:

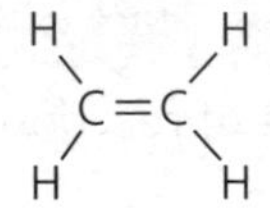

How are alkenes identified?
- Unlike alkanes, alkenes react with **aqueous bromine** (bromine water).
- Bromine water is **orange**. The product of the reaction is **colourless**.
- In the reaction, bromine atoms add on to the molecule at the double bond, in an **addition reaction**. This shows the reaction with ethene:

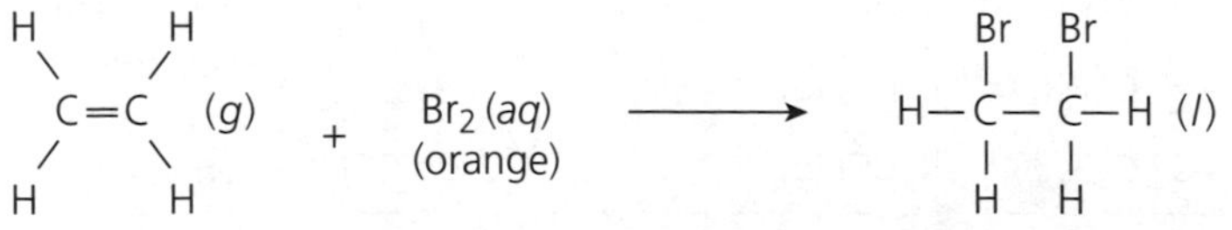

ethene (colourless) (g) + Br_2 (aq) (orange) → 1,2–dibromoethane (colourless) (l)

1,2-dibromoethane
What the name tells you:
- *dibromo* – has two bromine atoms
- *1,2* – they are on adjacent carbon atoms
- *ethane* – an alkane (so it is saturated)

- So the reaction occurs *because of* the double bond.
- Bromine water is therefore used as a test, to distinguish between saturated and unsaturated hydrocarbons. If a hydrocarbon gives a colour change from orange to colourless for bromine water, it is unsaturated.

Extended

More addition reactions of alkenes

In an addition reaction:

- a molecule **adds on** at the C=C of an alkene, making it into a single bond
- so the unsaturated molecule becomes **saturated**, and **no other product is formed**.

1 Addition reactions with hydrogen and steam

Reaction	unsaturated ⟶		saturated
With hydrogen	alkene	add on hydrogen, H_2	alkane
Example	ethene	$+ \ H_2$ (g) $\xrightarrow{\text{heat, pressure} \atop \text{a catalyst}}$	ethane
With steam	alkene	add on steam, H_2O	alcohol
Example	ethene	$+ \ H_2O$ (g) $\underset{\text{a catalyst}}{\overset{\text{heat, pressure}}{\rightleftharpoons}}$	ethanol

2 Polymerisation reactions

In these reactions:

- alkene molecules **add on to each other** in the presence of a catalyst, to form molecules with very long chains
- the compounds with these long-chain molecules are called **polymers**
- the alkene molecules are called **monomers**
- the process is called **addition polymerisation**.

Make the link to ...
addition polymerisation,
on page 154.

Reaction	from **alkene** (unsaturated) ⟶		**to poly(alkene)** (saturated)
Example	n **ethene** (the monomer)	$\xrightarrow{\sim 50°C, \ 3 \text{ or } 4 \text{ atmospheres pressure,} \atop \text{a catalyst}}$	**poly(ethene)** (the polymer)

Summary: comparing alkanes and alkenes

	Alkanes	Alkenes
Main source	fractional distillation of petroleum	cracking alkanes
Bonding	all single bonds	one carbon–carbon double bond
Type of compound	saturated hydrocarbon	unsaturated hydrocarbon
Characteristic reactions	substitution (requires UV light)	addition
With bromine water	no change – bromine water stays orange	orange to colourless
Polymerisation	do not form polymers	form polymers by addition
Example	ethane, C_2H_6	ethene, C_2H_4

Quick check for 20.5   *(Answers on page 170)*

1   Write the formula for the alkene with 5 carbon atoms.
2   Are alkenes generally *reactive*, or *unreactive*? Why?
3   Which test will distinguish between saturated and unsaturated hydrocarbons?
4   Why is the reaction between an alkene and bromine described as an *addition*?
5   Name the polymer obtained from the monomer propene.
6   How many products are there, in an addition reaction?

20.6 Alcohols

The alcohols are a family of organic compounds.

*   They contain the **OH** functional group.
*   Their names end in **-ol**.
*   They form a homologous series with the general formula $C_nH_{2n+1}OH$.
*   All their carbon atoms form **single bonds** so they are **saturated** compounds.

The most important alcohol is **ethanol, C_2H_5OH**.

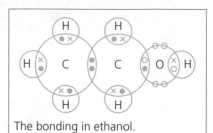

The bonding in ethanol.

How ethanol is obtained

Ethanol is obtained in two ways:

---

1   By fermentation of glucose – the 'biological' way

$$C_6H_{12}O_6\ (aq) \xrightarrow[\text{around 35°C}]{\text{enzymes in yeast}} 2C_2H_5OH\ (aq)\quad +\quad 2CO_2\ (g)\quad +\quad \text{energy}$$

glucose   ethanol   carbon dioxide

*   During fermentation, the **enzymes** in living **yeast cells** catalyse the breakdown of glucose into ethanol and carbon dioxide
*   Fermentation is an **exothermic** reaction – heat is released.
*   The ethanol is separated from the mixture by **fractional distillation** (page 9).

Fermentation is used to make ethanol from corn, and wine from grapes. When the % of ethanol reaches a certain level, or if the liquid gets too warm, the yeast stops working.

---

2   By hydration of ethene – the 'chemical' way

$$\begin{array}{c} H \quad\quad H \\ \diagdown\quad\diagup \\ C=C \quad (g) \\ \diagup\quad\diagdown \\ H \quad\quad H \end{array} \ + \ H_2O\ (g) \xrightleftharpoons[\substack{\text{a catalyst} \\ \text{(phosphoric acid)}}]{570°C,\ 60–70\ \text{atm}} \ \begin{array}{c} H \ \ H \\ | \ \ | \\ H-C-C-H\ (l) \\ | \ \ | \\ H \ \ OH \end{array} \ + \ \text{energy}$$

ethene   ethanol

*   The reaction is called a **hydration** because a **water molecule** adds on to the ethene.
*   There is no other product – so this is an **addition reaction.**
*   A catalyst is needed, to speed up the reaction.
*   It is a **reversible reaction**, and high pressure improves the yield of ethanol.
*   The hydration is **exothermic** – heat is released.
*   A low temperature will improve the yield, but the reaction will be too slow – so 570 °C is chosen as a compromise.

How ethanol is used
- Ethanol is the alcohol in alcoholic drinks.
- It is a good **solvent**. It dissolves many things that don't dissolve in water.
- It evaporates easily, which makes it a suitable solvent for cosmetics and glues.
- It is used as a **fuel**. It burns readily in air giving out plenty of heat. (In many countries it is mixed with petrol, or used on its own, for cars.) This is the equation for its complete combustion, in a plentiful suppply of air:

$$C_2H_5OH\ (l)\ +\ 3O_2\ (g)\ \longrightarrow\ 2CO_2\ (g)\ +\ 3H_2O\ (l)\ +\ heat$$

Note that in complete combustion, carbon dioxide and water are the only products formed.

✓

Quick check for 20.6   *(Answers on page 170)*
1. Write the formula of the alcohol which has 5 carbon atoms.
2. What is the functional group in an alcohol?
3. Why is the hydration of ethene called an *addition reaction*?
4. Write a word equation for the hydration of ethene.
5. Write the equation for the complete combustion of methanol (CH_3OH).

20.7 Carboxylic acids

The carboxylic acids are a family of organic compounds.
- They contain the **COOH** functional group, which has a **C=O double bond** in it.
- Their names end in **-oic acid**.
- They form the homologous series with the general formula $C_nH_{2n}O_2$.

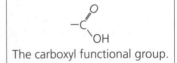
The carboxyl functional group.

The most common carboxylic acid is **ethanoic acid, CH_3COOH**.

How ethanoic acid is obtained
Ethanoic acid is produced by the oxidation of ethanol:

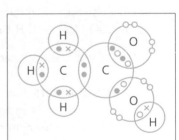
The bonding in ethanoic acid. Note the C=O double bond.

ethanol   →[O]→   ethanoic acid

| 1 | By fermentation – the 'biological' way |

- If ethanol is left standing in air, **bacteria** will bring about its oxidation to ethanoic acid.
- This process is called **acid fermentation**. It is used to make vinegar from fruits and other foodstuffs. (Vinegar is a dilute solution of ethanoic acid.)

| 2 | Using oxidising agents – the 'chemical' way |

- Ethanol is oxidised much faster by warming it with potassium manganate(VII), a powerful **oxidising agent,** in the presence of acid.
- The manganate(VII) ions are reduced to Mn^{2+} ions, with a colour change. The acid provides the hydrogen ions for this reaction:

$$MnO_4^-\ +\ 8H^+\ +\ 5e^-\ \longrightarrow\ Mn^{2+}\ +\ 4H_2O$$
purple   colourless

Using another oxidising agent

Ethanol is also oxidised by potassium dichromate(VI), another powerful oxidising agent, in the presence of acid. Again there is a colour change, when the chromate(VI) ions are reduced:

$$Cr_2O_7{}^{2-} + 14H^+ + 6e^- \longrightarrow 2Cr^{3+} + 7H_2O$$
orange   green

Potassium dichromate is used in one type of **breathalyser**. It oxidises any alcohol on a person's breath and is itself reduced, changing colour from orange to green.

Reactions of ethanoic acid

- Ethanoic acid is a **weak** acid. That means it partly ionises (or dissociates) in water, to form ethanoate ions and hydrogen ions:

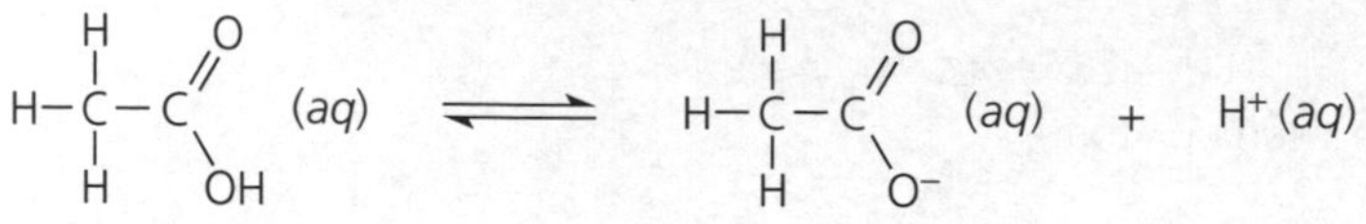

ethanoic acid   ethanoate ion   hydrogen ion

or CH_3COOH (aq) $\rightleftharpoons$ CH_3COO^- (aq) $+ H^+$ (aq)

- Being an acid, it neutralises bases, to form **salts** such as sodium ethanoate:

ethanoic acid   sodium   sodium ethanoate   water
                   hydroxide   (a salt)

or CH_3COOH (aq) $+$ NaOH (aq) $\longrightarrow$ CH_3COONa (aq) $+ H_2O$ (l)

- Ethanoic acid reacts with alcohols to form compounds called **esters**. For example with propan-1-ol it forms the ester **propyl ethanoate**:

ethanoic acid   +   propan-1-ol $\longrightarrow$ propyl ethanoate   +   water
CH_3COOH (aq) $+$ C_3H_7OH (aq) $\longrightarrow$ $CH_3COOC_3H_7$ (aq) $+ H_2O$ (l)

Note that the 'alcohol' part comes first in the name.

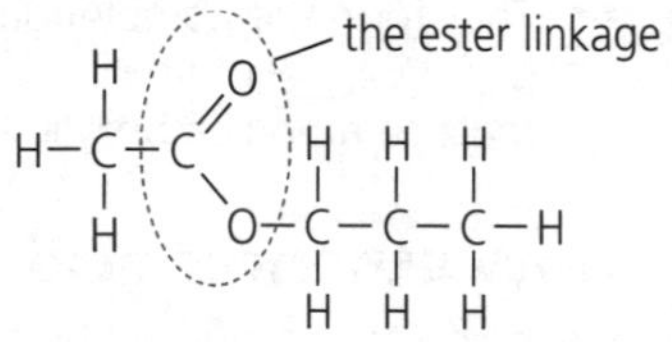

More about esters

- Esters are formed when an alcohol reacts with a carboxylic acid.
- The diagram on the right shows the structure of the ester propyl ethanoate. The functional group is circled. It is called the **ester linkage**.
- Many esters have distinctive tastes and smells, so they are used as artificial flavorings and fragrances.

Propyl ethanoate: the part from the acid is usually put first in the drawing, but second in the name.

✓

Quick check for 20.7   *(Answers on page 170)*

1. What is the functional group in a carboxylic acid?
2. Name the product that forms, when the alcohol methanol is oxidised.
3. How is the ethanoic acid in vinegar made?
4. Why does the colour change, when ethanol is oxidised by acidified potassium dichromate(VI)?
5. How could you tell that ethanoic acid is a *weak* acid?
6. How is an ester made?

Questions on section 20

Answers for these questions are on page 170.

Core curriculum

1   The compound shown on the right is the first member of the alkane homologous series.

$$\begin{array}{c} H \\ | \\ H-C-H \\ | \\ H \end{array}$$

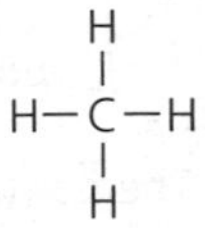

   a   State two characteristics of a homologous series.
   b   Name and draw the structure of the next member of the alkane homologous series.
   c   Complete the table to show the structure and uses of some organic compounds.

Name of compound	Molecular formula	Structure (showing all atoms and bonds)	Use
ethene	C_2H_4	i	ii
ethanoic acid	$C_2H_4O_2$	iii	making esters
dibromoethane	iv	$\begin{array}{c} Br\ \ Br \\ |\ \ \ \ | \\ H-C-C-H \\ |\ \ \ \ | \\ H\ \ \ H \end{array}$	
v	CH_4	$\begin{array}{c} H \\ | \\ H-C-H \\ | \\ H \end{array}$	vi

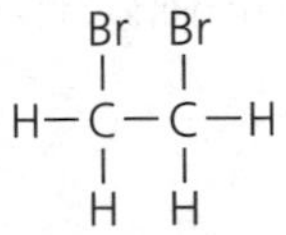

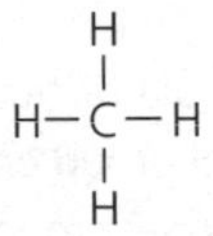

CIE 0620 November '08 Paper 2 Q6

2   Petrol is a mixture of alkanes.
   One of the alkanes in petrol is octane, C_8H_{18}.
   a   What products are formed when octane is completely burnt in air?
   b   More petrol can be made by cracking less useful petroleum fractions.
      i   What do you understand by the term *cracking*?
      ii   State two conditions needed for cracking.
      iii   Alkenes can be formed by cracking. The simplest alkene is ethene.
         Draw a diagram to show the structure of ethene.
         Show all atoms and bonds.
      iv   Complete the equation for the cracking of tetradecane, $C_{14}H_{30}$.
         $C_{14}H_{30} \longrightarrow$ .... $+ C_2H_4$

CIE 0620 June '07 Paper 2 Q3

3   Some reactions of organic compounds are shown below.
   A $nCH_2{=}CH_2 \longrightarrow +CH_2-CH_2+_n$

   B $C_3H_8 + 2O_2 \longrightarrow 3CO_2 + 4H_2O$

   C $C_6H_{12}O_6 \longrightarrow 2CO_2 + 2C_2H_5OH$

      glucose

   D $C_8H_{18} \longrightarrow C_6H_{14} + C_2H_4$

   a   i   Which **one** of the reactions, A, B, C, or D shows fermentation?
      ii   Which **one** of the reactions, A, B, C, or D shows polymerisation?
      iii   Which **one** of the reactions, A, B, C, or D shows combustion?
      iv   Which **one** of the reactions, A, B, C, or D shows cracking?
   b   The hydrocarbon C_8H_{18} is an alkane.
      i   What is meant by the term *hydrocarbon*?
      ii   Explain why this hydrocarbon is an alkane.

CIE 0620 June '04 Paper 2 Q4

4 The diagram shows a bottle of mineral water.

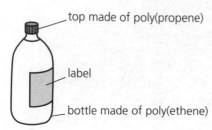

top made of poly(propene)

label

bottle made of poly(ethene)

a The poly(propene) top is made by polymerising propene molecules, $CH_3CH=CH_2$.
  i Which of the following best describes the propene molecules in this reaction?
    alkanes   monomers   polymers   products   salts
  ii State the name of the homologous series to which propene belongs.
  iii Propene is an unsaturated hydrocarbon.
    State the meaning of the terms *unsaturated* and *hydrocarbon*.
  iv Describe a chemical test to distinguish between an unsaturated hydrocarbon
    and a saturated hydrocarbon.
b The poly(ethene) bottle is made by polymerising ethene.

$$nCH_2=CH_2 \longrightarrow -(CH_2-CH_2)_n$$

What type of polymerisation is this?

CIE 0620 November '07 Paper 2 Q2

5 a Many sweets contain citric acid. The formula of citric acid is shown on the right.
  i Which group is the alcohol functional group on this formula?
  ii State the name of the $-CO_2H$ functional group in citric acid.
  iii Ethanoic acid also has a $-CO_2H$ functional group.
    Write down the formula for ethanoic acid.
b Nowadays, citric acid is usually made by the fermentation of sugars.
  Which one of the following is required for fermentation?
  acid   high temperature   light   microorganisms   nitrogen

CIE 0620 November '07 Paper 2 Q4

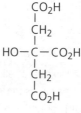

Extended

Extended curriculum

1 For each of the following, predict the name of the organic product.
  a reaction between methanol and ethanoic acid
  b oxidation of propan-1-ol by potassium dichromate (VI)
  c removal of H_2O from ethanol (dehydration)

CIE 0620 November '07 Paper 3 Q2

2 The fermentation of glucose is catalysed by enzymes from yeast. Yeast is added to
aqueous glucose, the solution starts to bubble, and becomes cloudy as more yeast cells
are formed.

$$C_6H_{12}O_6 \ (aq) \longrightarrow 2C_2H_5OH \ (aq) + 2CO_2 \ (g)$$

The reaction is exothermic. Eventually the fermentation stops when the concentration
of ethanol is about 12%.

a What is an enzyme?
b Pasteur said that fermentation was respiration in the absence of air. Suggest a
  definition of respiration.
c On a large scale, the reaction mixture is cooled. Suggest why this is necessary.
d Why does the fermentation stop? Suggest two reasons.
e When the fermentation stops, there is a mixture of aqueous ethanol and yeast.
  Suggest a technique that could be used to remove the cloudiness due to the yeast.
f Name a technique which will separate the ethanol from the ethanol / water
  mixture.

CIE 0620 June '08 Paper 3 Q6

Extended

3 The alkanes are generally unreactive. Their reactions include combustion, substitution and cracking.

a The complete combustion of an alkane gives carbon dioxide and water.

    i 10 cm³ of butane is mixed with 100 cm³ of oxygen, which is an excess. The mixture is ignited. What is the volume of unreacted oxygen left and what is the volume of carbon dioxide formed?

$$C_4H_{10}\ (g)\ +\ 6\tfrac{1}{2}\,O_2\ (g)\ \longrightarrow\ 4CO_2\ (g)\ +\ 5H_2O\ (l)$$

    ii Why is the incomplete combustion of any alkane dangerous, particularly in an enclosed space?

b The equation for a substitution reaction of butane is given below.

$$CH_3–CH_2–CH_2–CH_3\ +\ Cl_2\ \longrightarrow\ CH_3–CH_2–CH_2–CH_2–Cl\ +\ HCl$$

    i Name the organic product.

    ii This reaction does not need increased temperature or pressure. What is the essential reaction condition?

    iii Write a different equation for a substitution reaction between butane and chlorine.

c Alkenes are more reactive and industrially more useful than alkanes. They are made by cracking alkanes.

$$C_7H_{16}\ \longrightarrow\ CH_3–CH=CH_2\ +\ CH_3–CH_2–CH=CH_2\ +\ H_2$$
heptane   propene   but–1–ene

    i Draw the structural formula of the polymer poly(propene).

    ii Give the structural formula and name of the alcohol formed when but-1-ene reacts with steam.

    iii Deduce the structural formula of the product formed when propene reacts with hydrogen chloride.

CIE 0620 November '08 Paper 3 Q7

Alternative to practical

1 Ethene gas was formed by the cracking of a liquid alkane. The diagram shows the apparatus used.

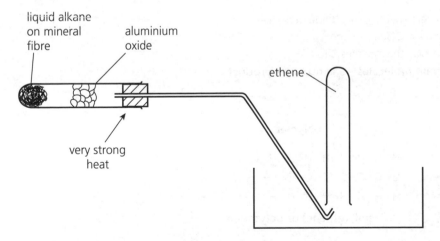

liquid alkane on mineral fibre   aluminium oxide   ethene

very strong heat

a Identify two mistakes in the diagram.

b Describe a test to show the presence of ethene. Give the result of the test.

CIE 0620 November '08 Paper 6 Q3

21 Synthetic polymers

The big picture

- Polymers are substances made of very large molecules, built up from small ones.
- There are hundreds of natural polymers. (For example the proteins in your hair and bones.) But we also make hundreds of synthetic polymers, including plastics.
- Plastics are very useful materials – but they can pollute the environment.

21.1 What are synthetic polymers?

Polymers

- Polymers are substances made of very large molecules, or **macromolecules**, which have been built up from small molecules called **monomers**.
- The reaction in which the large molecules are formed is called **polymerisation**.
- In most polymers, the macromolecules are based on long chains of carbon atoms.
- Polymers can be **natural** (such as the protein keratin, in your hair), or **synthetic**.

Synthetic polymers

- Synthetic polymers include polythene, PVC, and all the other materials we call **plastics**.
- They are made by two kinds of polymerisation reactions: **addition polymerisation** and **condensation polymerisation.**

21.2 Addition polymerisation

In addition polymerisation, the monomers are identical **alkene** molecules.

- Since they are alkenes, they have a C=C double bond.
- The breaking of this C=C bond allows the monomers to join.
- The polymerisation gives **long-chain molecules** – and **no other product**.

Example 1

monomer   polymer

ethene   poly(ethene) or polythene

Chains of different lengths form – but all are long, with thousands of carbon atoms. So we write the equation for the reaction like this, where n stands for a large number:

By changing the reaction conditions we can make different types of polythene.

Properties of polythene: light, bends without breaking, tough, water resistant.
Used to make: plastic bags, plastic bottles and cartons, toys, tables and chairs, water pipes, washing-up bowls.

Extended

Example 2

monomer

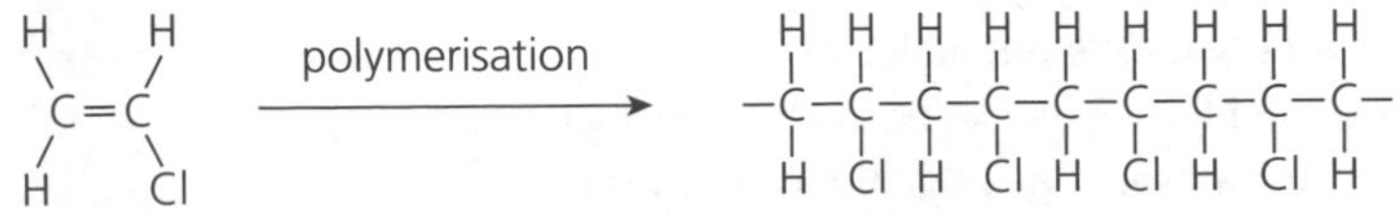

chloroethene
(vinyl chloride)

polymer

poly(chloroethene)
(polyvinyl chloride or PVC)

The equation for the reaction can be written like this, where n stands for a large number:

Properties of PVC: flexible, a good insulator, easily moulded, water resistant.

Used to make: waterproof clothing, covering for electrical wiring, dustbins, tables and chairs, water pipes.

Drawing the monomers and polymers

Going from monomer to polymer	Identify the monomer, from the polymer
• Draw the monomer units with their double bonds next to each other. • Replace the double bonds with single bonds, and add a new single carbon–carbon bond between the units.	• Identify the repeating unit. (It will have two carbon atoms joined by a single bond.) Put brackets around it. • Now draw this unit as a separate molecule with a **double bond** between the carbon atoms.
monomer phenylethene (styrene) **polymer** poly(phenylethene) poly(styrene)	**polymer** poly(propene) **monomer** propene

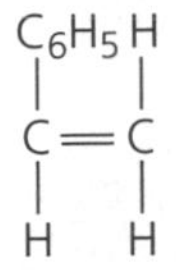

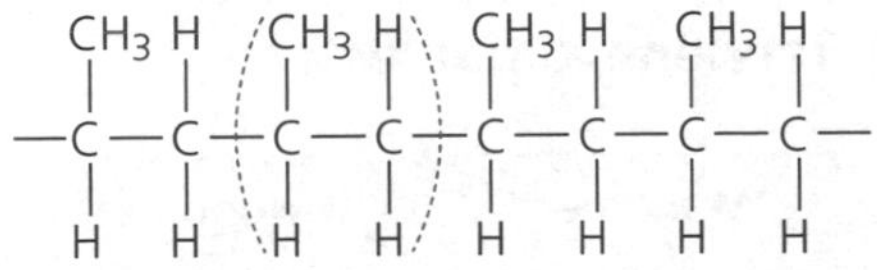

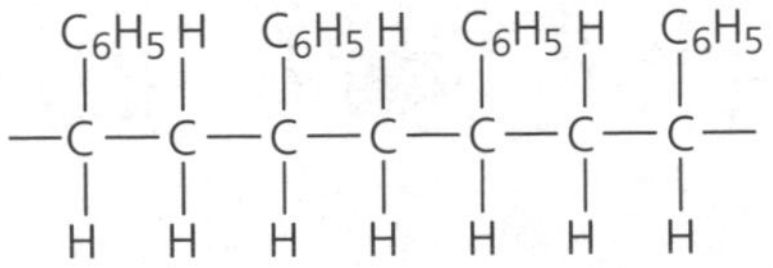

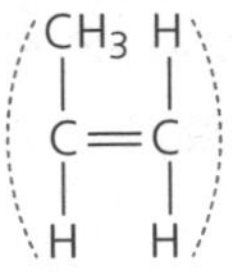

✓ **Quick check for 21.2** *(Answers on page 170)*

1 The reaction that gives polythene is called an *addition polymerisation*. Explain why. (You will need to explain both parts of the term.)

2 Can propane (C_3H_8) polymerise to form a polymer? Explain your answer.

3 Draw and name:

a the polymer made from this monomer, 2-propenamide

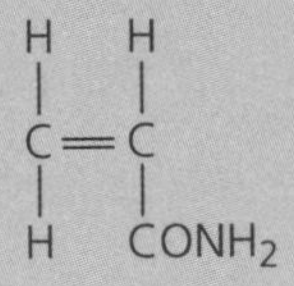

b the monomer used to make this polymer, poly(tetrafluoroethene)

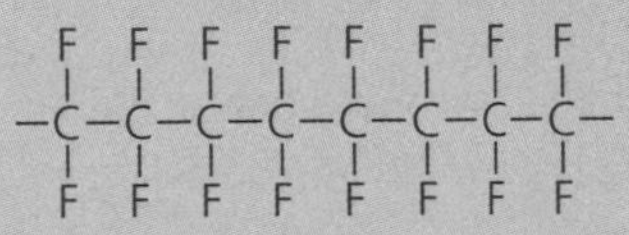

part of a polymer molecule

Extended

21.3 Condensation polymerisation

- In condensation polymerisation, the monomers are two *different* molecules.
- The polymerisation gives **long-chain molecules** – and there is always **one other product**.
- This other product is a small molecule which is **eliminated** when the monomers join.

Nylon and Terylene are examples of polymers made by condensation polymerisation. In the reactions shown below, blocks like —▢— and —▮— represent different carbon chains.

Nylon, a polyamide

Monomer A Monomer B	Polymer	Small molecule eliminated
diaminoalkane **di**oylchloride	amide linkage nylon – a polyamide	HCl, hydrogen chloride
Properties of nylon: can be drawn out into tough strong fibres that do not rot away. **Uses to make:** thread, ropes, fishing nets, carpets.		

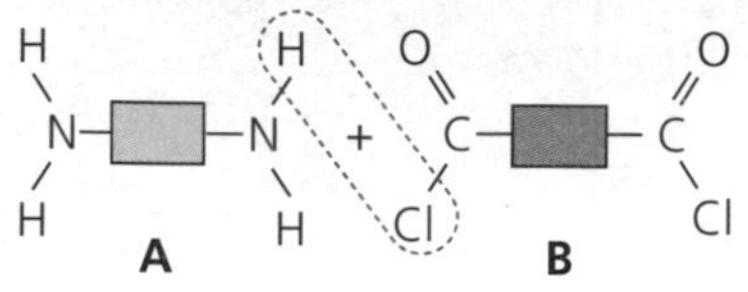

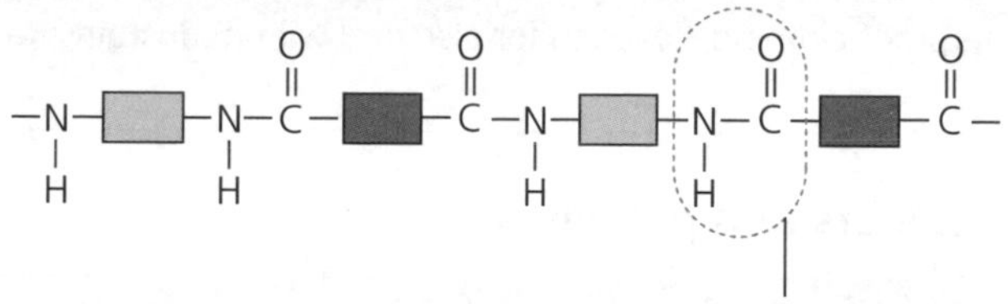

Terylene, a polyester

Monomer C Monomer D	Polymer	Small molecule eliminated
dicarboxylic acid **di**alcohol	ester linkage Terylene - a polyester	H$_2$O, water
Properties of Terylene: can be drawn out into tough light, hard-wearing fibre that is easily woven. **Uses to make:** thread, fabric for shirts and bed linen (often woven with cotton).		

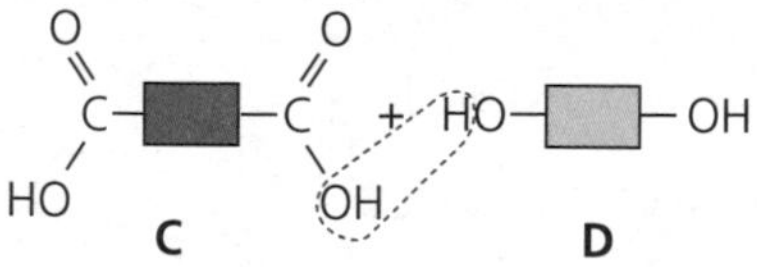

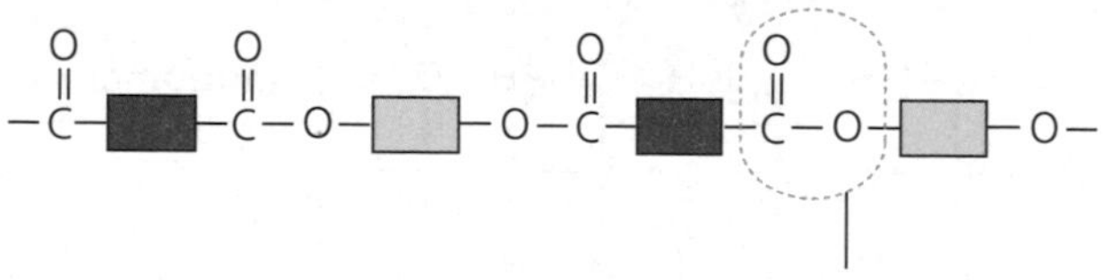

✓

Quick check for 21.3   (Answers on page 170)

1 In what way(s) is condensation polymerisation different from addition polymerisation?
2 Draw: **a** the amide linkage   **b** the ester linkage
3 **a** Which types of chemicals are required, to form a polyamide?
   b Describe how the chemicals react, to form the amide link.
4 Show how the formation of a polyester is: **a** similar to   **b** different from the formation of polyamide.

Extended

21.4 Plastic pollution

All the materials we call **plastics** are synthetic polymers, made by polymerisation reactions in industry. The starting compounds are very often obtained from the refining of petroleum.

Plastics are really useful materials. But they also pose a big problem.
- They do not rot away. They cannot be broken down by bacteria or other natural organisms – they are not **bio-degradeable**.
- So when we throw out plastic things, they become pollutants.

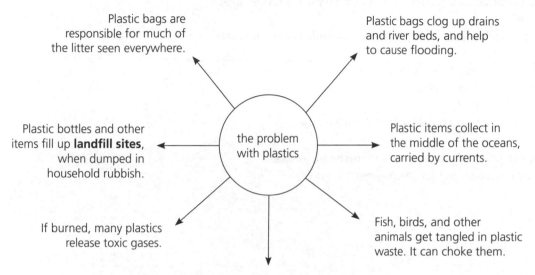

Plastic bags are responsible for much of the litter seen everywhere.

Plastic bags clog up drains and river beds, and help to cause flooding.

Plastic bottles and other items fill up **landfill sites**, when dumped in household rubbish.

the problem with plastics

Plastic items collect in the middle of the oceans, carried by currents.

If burned, many plastics release toxic gases.

Fish, birds, and other animals get tangled in plastic waste. It can choke them.

When fish, birds, and other animals swallow plastic, they cannot digest it. But it fills their stomachs, and they die of hunger.

Plastic bags are the worst problem, since billions are made every year, used once, and thrown away.

Tackling the problem
- Many countries have now banned plastic bags.
- In some countries, plastic materials are collected for recycling.
- Chemists are trying to develop bio-degradeable plastics, that will break down in the natural environment. Semi-degradeable plastics already exist. These break down into small flakes over time, when exposed to heat and moisture, or bacteria.

✓

Quick check for 21.4   *(Answers on page 170)*
1 Give two ways in which plastics can harm living things.
2 Give two other negative effects of plastics on the environment.
3 Would bio-degradeable plastics overcome these problems?

Questions on section 21

Answers for these questions are on page 170.

Extended curriculum

1   The polymer poly(propene) is used in the manufacture of ropes and textiles.
    a   Give the structural formula of the monomer.
    b   Draw the structural formula of the polymer.
    c   Which of the two structures, the monomer or the polymer, is classed as a macromolecule?
    d   Of the monomer and polymer, which one will react with aqueous bromine? Explain your answer.
    e   Name another use for polymers like poly(propene).
    f   Poly(propene) is not biodegradable.
        i   What does this mean?
        ii   Give two environmental problems that this may cause.

2   The addition polymer, polyacrylonitrile, is used to make carbon fibres.
    The diagram shows the repeat unit for this polymer:

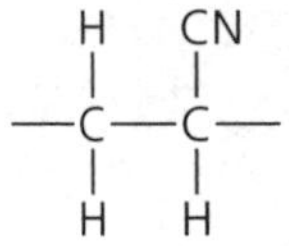

$$\begin{array}{cc} H & CN \\ | & | \\ -C- & C- \\ | & | \\ H & H \end{array}$$

    a   Copy the diagram and add two more units to the polymer chain.
    b   Name the monomer for this addition polymer.
    c   Draw the structure of the monomer.
    d   Write an equation for the polymerisation.
    e   Explain why is an example of addition polymerisation.

3   The fibres nylon and Terylene are both made by condensation polymerisation.
    a   What is *condensation polymerisation*?
    b   Which of the fibres is: i   a polyamide?   ii   a polyester?
    c   Give one use for each of these fibres.
    d   i   Which type of condensation polymer will be formed from these two monomers?
                $HO-CH_2-CH_2-OH$ and HO_2C-CO_2H
        ii   Draw the link that forms between the monomers.

22 Natural macromolecules

The big picture

- Many natural substances exist as macromolecules – giant molecules.
- They are natural polymers, built up from small units by polymerisation.
- The carbohydrates, proteins and fats in our food are natural polymers.

22.1 Proteins

- Proteins are natural polymers that contain the **amide linkage** (like nylon does).
- All living things contain proteins (including all the plants and animals we use for food).

Proteins		
The starting units	**amino acids** These are carboxylic acids, with an amino group (NH_2) on the carbon atom next to the COOH group.	R stands for the rest of the molecule. For example: – in glycine, R is H – in alanine, R is CH_3
How they join	By condensation polymerisation, with the elimination of water molecules. **H from the NH_2 group** of one molecule combines with **OH from the COOH group** of another molecule, to form a water molecule. A new C—N link forms.	
Example **From amino acids (R groups can vary)** **to the polymer – the protein chain**	water molecules eliminated amide linkage Many thousands of amino acid units can join together in this way.	
Breaking down and building up again	• During digestion the proteins in our food are **hydrolysed** back to amino acids. **Hydrolysis** means **the breaking down of a compound by reaction with water**. • Enzymes in the body act as catalysts for this. • The amino acids then join up again to make the proteins our bodies need.	
Why proteins are an essential part of our diet	• We must eat proteins to obtain the amino acids to make the proteins in hair, muscle, skin, and bone. • The enzymes in our bodies (which are catalysts for body reactions), and haemoglobin (the red substance in blood, that carries oxygen) are also proteins.	

✓

Quick check for 22.1   *(Answers on page 170)*

1   Which synthetic polymer has the same linkage as proteins?
2   What happens during the hydrolysis of a protein?

Extended

22.2 Carbohydrates

- Carbohydrates contain carbon, hydrogen and oxygen.
- The simplest carbohydrates are sugars such as glucose, which plants make. These are called **monosaccharides** (which means *single sugar units*).
- These units join to form macromolecules called **complex carbohydrates** or **polysaccharides**. An example is starch, which is found in rice and many other foods.

A molecule of glucose.

Carbohydrates	
The starting units	sugar units called **monosaccharides** – for example glucose, shown in the drawing above. We draw the sugar units simply, like this, with just two OH groups shown: HO—▢—OH
How they join	By **condensation polymerisation**, with the elimination of **water molecules**. **H from the OH group** on one molecule combines with an **OH** from another molecule, to form a water molecule. A new **C-O** link forms.
Example: from a simple carbohydrate such as glucose ↓ **to a complex one** such as starch	HO—▢—OH HO—▢—OH HO—▢—OH HO—▢—OH water molecules eliminated —O—▢—O—▢—O—▢—O—▢— Up to 2500 glucose units can join together in this way.
Breaking down and building up again	• During digestion the starch in our food is **hydrolysed** back to glucose. • Enzymes act as catalysts for this. • Some glucose is then used to form another complex carbohydrate, glycogen.
Why carbohydrates are an essential part of our diet	• Glucose provides energy, through the process called **respiration**, which takes place in all our body cells (page 127). • Glycogen is used as an energy store for the body. (It is stored in the liver, and in muscle cells.)

✓ **Quick check for 22.2** *(Answers on page 170)*
1 Describe how a complex carbohydrate is made from sugar units.
2 Draw part of a carbohydrate chain, containing 4 sugar units.

22.3 Fats

- Fats are **esters**, made from the reaction between carboxylic acids, and an alcohol. So they contain the ester linkage – like the polyester Terylene does.
- The carboxylic acids used are natural acids with long chains. They are called **fatty acids**.

Fats		
The starting units	**fatty acids** (R represents a long chain) R—C(=O)—OH and	**glycerol** (an alcohol with 3 OH groups) HO—CH₂ HO—CH HO—CH₂
How they join	By **condensation polymerisation**, with the elimination of **water molecules**. **Three OH** from three acid molecules combine with the **three H** from one glycerol molecule. Three new C—O links form.	

Extended

The condensation reaction	

3 molecules of fatty acid   1 molecule of glycerol   1 macromolecule of a fat

Breaking down	• During digestion, fats are **hydrolysed** back to glycerol and fatty acids. • Enzymes act as catalysts for this. • The glycerol is converted to glucose by the liver.
Why fats are an essential part of our diet	Fats are used to make cell membranes, insulate our bodies, maintain healthy skin and hair, and to provide an energy store for the body.

✓ **Quick check for 22.3**   *(Answers on page 170)*
1 Which macromolecule has the same linkage as Terylene?
2 What use does your body make of the products from the hydrolysis of fats?

Remember
Proteins, carbohydrates and fats all form by condensation polymerisation, with the elimination of water molecules.

22.4 Hydrolysis in the lab

Proteins, complex carbohydrates, and fats are broken down by **hydrolysis** in your body – and this can also be carried out in the lab. Hydrolysis is the *reverse* of the condensation reactions that formed these compounds. The water is put back, and the macromolecules are broken down into smaller units.

Macromolecule	Conditions for the hydrolysis	Result of the hydrolysis
Proteins	boil with 6M hydrochloric acid for 24 hours	individual amino acids
Complex carbohydrates (eg starch)	heat with dilute hydrochloric acid	individual simple sugar units
Fats (esters of glycerol and fatty acids)	boil with dilute sodium hydroxide	glycerol + **sodium salts** of the fatty acids ($R\text{-}COO^- Na^+$)

• The individual amino acids and simple sugars can be **separated** and **identified** using **paper chromatography**.
• The simple sugars can be further broken down into **ethanol** and **carbon dioxide** in a **fermentation reaction** (page 148). The enzymes in yeast are catalysts for this reaction.
• The sodium salts of the fatty acids ($R\text{-}COO^- Na^+$) are used as **soaps**.

Make the link to ... separating colourless substances, on page 11.

✓ **Quick check for 22.4**   *(Answers on page 170)*
1 a What happens during the hydrolysis of a protein?
   b How is this hydrolysis carried out in the lab?
2 What are the experimental conditions for the hydrolysis of starch?
3 How would you identify the amino acids in a protein? (Two stages are required for this!)

Questions on section 22

Answers for these questions are on page 170.

Extended

Extended curriculum

1 Esters, fats and polyesters all contain the ester linkage.

    a The structural formula of an ester is given below.

       Name two chemicals that could be used to make this ester and draw their structural formulae. Show all bonds.

    b **i** Draw the structural formula of a polyester such as Terylene.

       ii Suggest a use for this polymer.

Cooking products, fats and vegetable oils, are mixtures of saturated and unsaturated esters. The degree of unsaturation can be estimated by the following experiment. 4 drops of the oil are dissolved in 5 cm^3 of ethanol. Dilute bromine water is added a drop at a time until the brown colour no longer disappears. Enough bromine has been added to the sample to react with all the double bonds.

Cooking product	Mass of saturated fat in 100 g of product / g	Mass of unsaturated fat in 100 g of product / g	Number of drops of bromine water
margarine	35	35	5
butter	45	28	4
corn oil	15	70	10
lard	38	56	?

    c **i** What is the value for the number of drops of bromine water for lard ?

       ii Complete the equation for bromine reacting with a double bond.

       iii Using saturated fats in the diet is thought to be a major cause of heart disease. Which of the products is the least likely to cause heart disease?

    d A better way of measuring the degree of unsaturation is to find the iodine number of the unsaturated compound. This is the mass of iodine that reacts with all the double bonds in 100 g of the fat.

       Use the following information to calculate the number of double bonds in one molecule of the fat.

       Mass of one mole of the fat is 884 g.

       One mole of I$_2$ reacts with one mole of C=C bonds

       The iodine number of the fat is 86.2 g.

       M$_r$ for I$_2$ is 254.

       86.2 g of iodine reacts with 100 g of fat.

       i What **mass** of iodine reacts with 884 g of fat?

       ii How many **moles** of iodine **molecules** reacts with 1 mole of fat?

       iii How many double bonds are there in **one molecule** of fat?

CIE 0620 June '07 Paper 3 Q7

2 Large areas of the Amazon rain forest are cleared each year to grow soya beans. The trees are cut down and burnt.

a Why do these activities increase the percentage of carbon dioxide in the atmosphere?

b Soya beans contain all three main food groups, two of which are protein and carbohydrate.

   i What is the third group?

   ii Draw the structural formula of a complex carbohydrate such as starch.

c Compare the structure of a protein with that of a synthetic polyamide. The structure of a typical protein is given below.

$$-N-\boxed{}-C-N-\boxed{}-C-N-\boxed{}-C-N-\boxed{}-C-$$
$$\;\;\;\;|\qquad\qquad\|\;\;|\qquad\qquad\|\;\;|\qquad\qquad\|\;\;|\qquad\qquad\|$$
$$\;\;\;\;H\qquad\;\;\;O\;\;H\qquad\;\;O\;\;H\qquad\;\;O\;\;H\qquad\;\;O$$

   i How are they similar?

   ii How are they different?   *CIE 0620 June '08 Paper 3 Q8*

3 The three types of food are carbohydrates, proteins and fats.

a Aqueous starch is hydrolysed to maltose by the enzyme amylase. The formula of maltose is:

$$HO-\boxed{}-O-\boxed{}-OH$$

Starch is hydrolysed by dilute sulfuric acid to glucose.

$$HO-\boxed{}-OH$$

   i What is an enzyme?

   ii Draw the structure of starch.

   iii Name the technique that would show that the products of these two hydrolyses are different.

b Proteins have the same linkage as nylon but there is more than one monomer in the macromolecule.

   i Draw the structure of a protein.

   ii What class of compound is formed by the hydrolysis of proteins?

c Fats are esters. Some fats are saturated, others are unsaturated.

   i Write the word equation for the preparation of the ester, propyl ethanoate.

   ii Deduce the structural formula of this ester, showing each individual bond.

   iii How could you distinguish between these two fats?

   Fat 1 has the formula

$$CH_2-CO_2-C_{17}H_{33}$$
$$|$$
$$CH-CO_2-C_{17}H_{33}$$
$$|$$
$$CH_2-CO_2-C_{17}H_{33}$$

   Fat 2 has the formula

$$CH_2-CO_2-C_{17}H_{35}$$
$$|$$
$$CH-CO_2-C_{17}H_{35}$$
$$|$$
$$CH_2-CO_2-C_{17}H_{35}$$

   iv Both of these fats are hydrolysed by boiling with aqueous sodium hydroxide. What type of compounds are formed?

          CIE 0620 November '06 Paper 3 Q8

Extended

Answers to questions

Section 1
Quick checks
for 1.1 **1 a** because the water particles can slide past each other **b** because the water particles can't move any closer **2** because the gas particles in the air can be squeezed a lot closer **3** Diffusion: the sugar particles collide with water particles, bounce away, and in this way spread through the water.

for 1.2 **1 a** because the vibrations of the particles get larger **b** The particles will vibrate so much that the structure breaks down. **2 a** gas **b** 130 °C **3** The heating curve will be like the one on page 3, but with the horizontal sections at 130 °C (melting point) and 240 °C (boiling point).

for 1.3 **1** When they meet, they combine to form a white cloud of ammonium chloride, which is easy to see.
2 a chlorine, since it has the larger relative molecular mass **b** raise the temperature

Questions on section 1
Core curriculum **1 a** See the diagrams on page 6. **b i** They only vibrate. **ii** They move around, and slide past each other. **c** They come close together, with forces of attraction between the particles **d** the gas state **e** The particles move around faster, with more energy. They collide more often and bounce further away. **2 a** The crystal dissolves and the particles spread by diffusion: they collide with water molecules and bounce away. **b i** regular, with particles close together **ii** They only vibrate. **3 a** ammonia, hydrogen chloride, and the gases in air **b i** ammonium chloride **ii** white **c** It shows that gas particles (ammonia and hydrogen chloride) have spread from the ends of the tube, and meet at X, where they react. **d** They collide with gas particles in air and bounce away again. **e** Ammonia will diffuse all through the tube – but no white solid will form.
Extended curriculum **1 a i** melting **ii** boiling **b i** close together **ii** far apart **iii** random and fast **iv** no **v** yes
2 a i See page 1. **ii** non-existent **iii** They move freely and fast. **b i** hydrogen, helium, oxygen, chlorine **ii** The lower its relative molecular mass, the faster a gas will diffuse. **c** It will diffuse twice as fast as oxygen. **d** It is between 32 and 71.

Section 2
Quick checks
for 2.1 **1** weighing balance for the salt, measuring cylinder for water **2** pipette **3 a** 2500 g **b** 500 g **4 a** 500 cm³ **b** 2000 cm³ **5 a** 2 dm³ **b** 0.5 dm³ **c** 1.5 dm³ **d** 0.5 dm³
for 2.2 **1 a** It is impure. **b** below −114.3°C **2** food additives, medical drugs, vaccines
for 2.3 **1 a** the substance you dissolve **b** the liquid you dissolve the substance in **2** fractionating column and thermometer added for fractional distillation **3 a** to separate the liquid from any solution of your choice **b** to separate any mixture of two or more miscible liquids **4** As the crystals form, any impurities are left behind in the liquid.
for 2.4 **1** Each dye has different solubility in the solvent and attraction to the paper. So the dyes move at different speeds up the paper. **2** the yellow dye **3** Each amino acid has a unique R_f value in a given solvent. **4** as a single spot (showing up after a locating agent is used) **5** 0.5625

Questions on section 2
Core curriculum **1 a** simple distillation **b** fractional distillation **c** crystallisation **d** fractional distillation **e** filtration **2 a** Bacteria and soil particles are larger than the gaps in the limestone, but water particles are smaller. **b i** condenser **ii** in the beaker **iii** It is higher. **3 a** soluble **b** by evaporating the water, leaving

salt crystals **c** simple distillation **d** insoluble **4 a** Place a spot of the solution on a pencil base line on the paper, and stand the paper in a beaker containing a little solvent. **b** It contains two metals. **c** A **d** B
Extended curriculum **1 a** (ii) **b** B **c** Use a locating agent. **d** A, 0.65; B, 0.4 **e** Look up the R_f values of amino acids in the solvent **f** Amino acids have different solubilities in different solvents.
Alternative to practical **1 a** A, thermometer **B**, beaker **C**, tripod **D**, condenser **b** to cool and condense the vapour **c** Measure its boiling point. **2 a** to prevent ink spreading up the paper **b** 1 and 3 **c** 4; two spots **d** to show up the amino acids

Section 3
Quick checks
for 3.1 **1 a** neutron **b** electron **c** proton **d** electron **2 a** 3 **b** 4 **c** 7 **d** $_3^7$Li **3** The atoms are: **a** sodium, with 11p 11e 12n **b** calcium, with 20p 20e 20n **c** iron, with 26p 26e 30n
for 3.2 **1** atoms of the same element, with different numbers of neutrons **2 a** 6, 7, and 8 **b** $_6^{12}$C (called C-12) **c** $_6^{14}$C **3** It is unstable and decays, giving out radiation.
for 3.3 **1 b** (2 + 8 +1) **2** Show three shells, with 2 electrons in the inner shell, 8 in the second, and 6 in the outer shell. **3** Note: for the drawings in this question you can show electrons in pairs as on page 19.) **a** Show one shell with 1 electron in it. **b** Show two shells: 2 electrons in the inner shell, 1 in the second. **c** Show four shells: 2 electrons in the inner shell, 8 in the second and third shells, and 1 in the fourth shell.
for 3.4 **1 a i** lithium **ii** fluorine **iii** chlorine **b** fluorine and chlorine: both have 7 electrons in their outer shells **2** phosphorus, 2 + 8 + 5 **3 a** helium **b** phosphorus **4 a** helium **b** nitrogen **c** magnesium **d** chlorine **e** calcium
5 a 1 **b** 2 + 8 + 2 **c** 2 + 8 + 18 + 8 + 1

Questions on section 3
Core curriculum **1 a** chlorine, argon, potassium, bromine, iodine **b** chlorine, potassium, argon, bromine, iodine **c** B **d** chlorine, bromine, iodine **e i** potassium **ii** argon **2 a i** atoms of the same element, with different numbers of neutrons **ii** Br-79 has 35e, 44n, 35p; Br-81 has 35e, 46n, 35p
3 a electrons 1−, neutrons no charge, protons, 1+ **b** 56 **c** as tracers to detect leaks in oil and gas pipes; for sterilizing foodstuffs and other things
Extended curriculum **1 a** electrons: mass 1/1840 which is taken as zero, charge 1−; protons: symbol p, charge 1+; neutrons, mass 1 **b i** They have equal numbers of protons and electrons, so charges cancel. **ii** They can lose electrons, which leaves more protons than electrons, giving a positive charge. **iii** The atoms can have different numbers of neutrons. **iv** An element has already been discovered, for each proton number. **2 a** ^{40}Ar: 18p, 18e, 22n; ^{40}Ca: 20p, 20e, 20n; ^{44}Ca: 20p, 20e, 24n **b i** They have the same number of protons but different numbers of neutrons. **ii** They do not have the same number of protons. **c** + 8 + 2 **3 a** 2 protons, 2 electrons, and 2 neutrons more **b** + 8 + 2
4 a B is $_{11}^{23}$Na, C is $_{18}^{40}$Ar, D is $_{11}^{31}$P³⁻, E is $_{13}^{27}$Al³⁺
b B – the same number of protons

Section 4
Quick checks
for 4.1 **1** The iron and sulfur atoms in it are held together by chemical bonds. **2** looks different, the iron and sulfur atoms in it are chemically bonded, it cannot be separated into iron and

sulfur by physical means **3** See the table on page 23. **4** add substances to turn into an alloy

for 4.2 **1** bonding: how atoms are joined; structure: how the bonded atoms are arranged **2** ionic: electrons transferred; covalent: electrons shared **3** a regular arrangement of particles **4** molecule **5** a lattice of silver ions, Ag^+, in a sea of electrons

for 4.3 **1** loses one electron, because that leaves it with a full outer shell of 8 electrons, which is a stable arrangement **2** as for sodium and chlorine on page 25, but with one middle shell less for each atom **3** regular arrangement of alternating positive and negative ions held in a lattice **4 a** and **b**: as for magnesium oxide and magnesium chloride on pages 25 and 26, but calcium has an extra inner shell of 8 electrons **5 a** NaOH **b** $AlCl_3$ **c** $Mg(OH)_2$

for 4.4 **1** Each shares its one electron. **2** Each atom has gained an outer shell of 8 electrons (or 2 in H_2). **3** Your diagram should be like the diagram for methane on page 28, but with chlorine atoms (7 outer-shell electrons) in place of hydrogen atoms. **4** lattice with regular arrangement of hydrogen molecules **5** The forces between molecules are weak. **6 a** and **b**: see page 28.

7

for 4.5 **1** covalent bonds **2 a** Both have covalent bonding. **b** Molecules contain small numbers of atoms, macromolecules contain millions. **3** The weak forces between layers allow them to slide over each other, so graphite is soft and slippery; the free electrons allow it to conduct. **4** Similar bonding and structure to diamond, but different from that in graphite.

for 4.6 **1** It is the attraction between metal ions and the sea of free electrons. **2** It forms a giant regular lattice. **3** They contain free electrons that can move, carrying current.

Questions on section 4

Core curriculum **1** a Check your drawing against the diagram for H_2O on page 28. **b** P_2O_3 **2 i** C **ii** E **iii** F **iv** E **v** D **vi** B **vii** A, D and F **3 a** The layers of atoms can slide over each other, making it slippery. **b** Each atom is bonded to four others, with strong covalent bonds.

Extended curriculum **1**

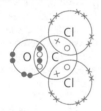

2 i good **ii** sodium chloride (or similar) **iii** silicon dioxide **iv** ions **v** electrons **vi** good **3 a** Your drawing should be like the one for $MgCl_2$, top right of page 26, but show only the outer shell of 8 electrons for the bromine ion. **b i** alternate positive and negative ions in a regular arrangement **ii** This ratio allows the total charges on the ions to balance.

Section 5
Quick checks
for 5.1 **1 a** carbon and hydrogen, ratio 1:4 **b** nitrogen and hydrogen, ratio 1:3 **c** iron and sulfur, ratio 1:1 **d** calcium, carbon and oxygen, ratio 1:1:3 **2** methane + oxygen $\longrightarrow$ carbon dioxide + water **3** 1:2:1:2
for 5.2 **1** $H_2(g) + Cl_2(g) \longrightarrow 2HCl(g)$
2 a $2Mg(s) + O_2(g) \longrightarrow 2MgO(s)$
b $4Fe(s) + 3O_2(g) \longrightarrow 2Fe_2O_3(s)$
3 $2H_2O_2(l) \longrightarrow 2H_2O(l) + O_2(g)$
for 5.3 **1** the mass of an atom compared to an atom of carbon-12 **2 a** 48 **b** titanium **3** 40 **4** It is the average value for two isotopes. **5 a** 17 **b** 32 **c** 106 **6** It is an ionic compound.

for 5.4 **1** 100 g **2 a** 160 g **b** 180 g **c** 400 g
3 a copper(II) carbonate $\longrightarrow$ copper(II) oxide + carbon dioxide **b** copper(II) oxide 20 g, carbon dioxide 11 g

Questions on section 5
Core curriculum **1 a** $2H_2 + O_2 \longrightarrow 2H_2O$
b $N_2 + 3H_2 \longrightarrow 2NH_3$ **c** $I_2 + Cl_2 \longrightarrow 2ICl$
d $2P + 3Cl_2 \longrightarrow 2PCl_3$ **2 a** 6 g **b** 97.5% **3 a** 169 **b i** $XeOF_4$ or XeF_4O **ii** 223 **iii** covalent with one double bond Xe=O
4 a $2Mg(s) + O_2(g) \longrightarrow 2MgO(s)$ **b** 40 **c i** 4.0 g **ii** 1.6 g
5 a 217 **b** 20.1 g of mercury, 1.6 g of oxygen

Section 6
Quick checks
for 6.1 **1** 6.02×10^{23} **2 a** 71 g **b** 35.1 g **3 a** 4 **b** 0.075 **c** 0.04
for 6.2 **1 a** CH **b** C_6H_6 **2** both NO_2
for 6.3 **1** 512 g **2** 12 g **3** 68 g **4** 4.76 g of MgO, 5.24 g of CO_2
for 6.4 **1 a** 0.125 mol/dm³ **b** 6 mol/dm³ **2 a** 4 dm³ **b** 0.02 dm³ or 20 cm³ **3 a** 0.15 **b** 0.1 **4 a** 112 g **b** 0.56 g **5** 0.1 dm³ or 100 cm³ **6** 0.015 dm³ or 15 cm³
for 6.5 **1 a** 4.5 dm³ or 4500 cm³ **b** 2.4 dm³ or 2400 cm³
2 1.2 dm³ or 1200 cm³ **3** 125 dm³ of both
for 6.6 **1** 100% **2** 75% **3** 80% **4** 18.2 tonnes **5** 83.3%

Questions on section 6
Extended curriculum **1 a** 0.3 g of $CaCO_3$ = 0.003 moles. That needs 0.006 moles of HCl for complete reaction. But 0.005 moles of HCl are used. So $CaCO_3$ is in excess. **b** volume of CO_2 = 0.06 dm³ or 60 cm³ **2 a** Repeat titration without the indicator, evaporate to remove some water, allow to cool to obtain crystals, dry the crystals. **b i** 0.056 **ii** 0.028 **iii** 9.016 g **iv** 42.8% **3 a** anhydrous copper(II) sulfate **b** powder turns blue **c** copper(II) oxide **d** 1.6 g; the equation is $CuSO_4 \longrightarrow CuO + SO_3$ so 1 mole of $CuSO_4$ gives 1 mole of CuO; M_r of $CuSO_4$ is 160; 5/250 x 80 = 1.6 g **4 a** See page 39. **b i** moles of Mg = 0.125; that needs 0.25 moles of CH_3COOH, or 15 g, for complete reaction. But there is only 12 g, so Mg is in excess. **ii** 0.1 moles (from 0.2 moles of acid) **iii** 2.4 dm³ or 2400 cm³ **c i** 0.01 moles **ii** 0.005 moles **iii** 0.25 mol/dm³ **5 a i** 0.08 **ii** 9.52 g **iii** 2.48 g **b i** 0.08 **ii** 22.48 g **iii** 46.3% **6** 1 (one)

Section 7
Quick checks
for 7.1 **1 a** and **e** Yes, metals contain electrons free to move. **b** Yes, ions in solution are free to move. **c** and **d** No, there are no ions in sugar. **2 a** ions not free to move **b** Melt it, or dissolve it in water.
for 7.2 **1 a** It carries current into and out of the liquid, during electrolysis. **b** the liquid which is electrolysed **2 a** connected to the positive terminal of the battery **b** anode **3** metal at cathode, non-metal at anode **4 a** potassium and chlorine **b** hydrogen and oxygen
for 7.3 **1** $Na^+(l) + e^- \longrightarrow Na(l)$; $2Cl^-(l) \longrightarrow Cl_2(g) + 2e^-$
3 A tiny % of water molecules is ionised to H^+ and OH^- ions.
for 7.4 **1** Copper ions are removed when copper forms at the cathode. **2 a** As copper ions are removed, more form at the anode. **b** The anode loses copper. For each atom it loses, one is deposited on the cathode.
for 7.5 **1** Use a nickel anode, the spoon as cathode, and a solution of a soluble nickel compound as electrolyte. **2** In both, metal atoms are lost from the anode, and deposited on the cathode. **3** A dilute solution would give oxygen, not chlorine. **4** by dissolving it in cryolite **5** $Al^{3+} + 3e^- \longrightarrow Al$; $2O^{2-} \longrightarrow O_2 + 4e^-$

Questions on section 7
Core curriculum **1 a** $PbBr_2$ **b** giant, ionic **c i** B **ii** platinum **iii** to make the ions free to move **iv** bromine at anode; lead at cathode **2 a** cathode **b** to be the electrolyte **c** chromium **3 a** cathode **b i** Copper is deposited on it. **ii** It dissolves.

Extended curriculum **1 a** Melt the compound, or dissolve it in water. **b** Positive ions move to the cathode, negative ions to the anode. **c i** $2Cl^- \longrightarrow Cl_2 + 2e^-$ **ii** $Cu^{2+} + 2e^- \longrightarrow Cu$ **d i** Oxygen forms rather than chlorine.
ii $4OH^- \longrightarrow 2H_2O + O_2 + 4e^-$ **e** Lithium is more reactive than hydrogen, copper is less reactive than hydrogen. **f i** It loses copper atoms as ions, so it dissolves. **ii** $Cu \longrightarrow Cu^{2+} + 2e^-$
iii oxidation **2 a** splitting by electricity **b** hydrogen ion H^+, chloride ion Cl^-, hydroxide ion OH^-
c i hydrogen: $2H^+ + 2e^- \longrightarrow H_2$ **ii** reduction: gain of electrons
d i oxygen **ii** chlorine **3 a i** $2H^+ + 2e^- \longrightarrow H_2$
ii $2Cl^- \longrightarrow Cl_2 + 2e^-$ **iii** Na^+ and OH^- ions are left in solution. **b i** It kills microbes/germs/bacteria. **ii** ammonia/hydrogen chloride/margarine **iii** sodium hydroxide **4 a** bauxite **b** to dissolve the aluminium oxide (melting point of aluminium oxide is too high) **c** It collects at the bottom of the tank. **d** graphite **e** A **f** It is above carbon in the reactivity series. **g i** from the aluminium oxide **ii** CO_2 **iii** The carbon reacts with the oxygen, forming carbon dioxide, a gas. **h** 530kg **i** (i) molten (ii) ions
Alternative to practical **1 a** A silvery metal appears at bottom of tube. **b** carbon or platinum **c** bromine, at the positive electrode (anode) **d** Bromine is poisonous.

Section 8
Quick checks
for 8.1 **1** given out **2** See the second graph on page 57.
3 magnesium oxide **4** It releases energy. **5 a** endothermic
b The energy to break bonds is greater than the energy released in forming bonds (the opposite of when ammonia is made).
for 8.2 **1** 92kJ

2 a

$$H-\underset{\underset{H}{|}}{\overset{\overset{H}{|}}{C}}-H + 2\ O=O \longrightarrow O=C=O + 2\ O\overset{\diagup H}{\diagdown H}$$

b 2648 kJ **c** 3466 kJ **d** exothermic **e** – 818 kJ
for 8.3 **1** Fossil fuels are burned to release energy; they do not give out radiation. Nuclear fuels are not burned, but give out radiation. **2** burning in oxygen **3** no harmful products **4** It would be no good as a fuel because its combustion is endothermic.
for 8.4 **1** two different metals standing in an electrolyte **2** from the more to the less reactive metal **3 a** zinc: $Zn^{2+} + 2e^- \longrightarrow Zn$
b It provides ions to carry the current. **4 a** 0.46 volts (= 0.78 – 0.32 volts) **b** lead **5** Hydrogen combines with oxygen to form water. **6** No greenhouse gases form.

Questions on section 8
Core curriculum **1 i** A **ii** A **iii** B **iv** B
2 a $2H_2(g) + O_2(g) \longrightarrow 2H_2O(l)$ **b** It can be obtained from water. **c** exothermic
Extended curriculum **1 a i** endothermic **ii** 3 **iii** endothermic **iv** N–H **v** exothermic **b** exothermic (75 kJ of energy out)
2 a i heat **ii** exothermic **iii** $C_2H_5OH\ (l) + 3O_2\ (g) \longrightarrow 2CO_2(g) + 3H_2O\ (l)$ **iv** like the first graph on page 57, with the formulae of reactants and products written in **b** around –2670 kJ/mol (The increase is –640 from methanol to ethanol and –650 from ethanol to propan-1-ol.) **3 a i** It will increase. **ii** zinc – more reactive **iii** The flow is from the more to the less reactive metal.
b In electrolysis, electrical energy causes chemical change; in cells, chemical change produces electrical energy. **4 a i** C–H or O=O **ii** C=O or O–H **iii** The energy released in making bonds is greater than the energy needed to break bonds. **b i** U –235 **ii** See page 17.
Alternative to practical **1** Use same volume or mass of each fuel, same starting water temperature, same burning time, same amount of stirring with thermometer. Record starting water temperature, burn first fuel for the planned time, record new water temperature. Repeat for second fuel. The fuel that gives the greater temperature rise gives out more energy.

Section 9
Quick checks
for 9.1 **1** how fast a reaction goes **2** how something like volume or mass changes over time **3** It is greatest at the start, and decreases over time until it reaches zero. **4** When the line becomes horizontal; that shows the rate is zero. **5** 15cm³/minute **6** Weigh the reaction flask at intervals. Its mass will decrease as carbon dioxide bubbles off.
for 9.2 **1 a** rate increases **b** rate decreases **2** powdered chalk **3 a** Both speed up the decomposition of hydrogen peroxide.
b The enzyme is made by living cells, and works in a limited range of temperature and pH.
for 9.3 **1** that the rate of a reaction depends on the number of successful collisions between reactant particles **2** Particles move more slowly, with less energy, so there are fewer collisions, and fewer are successful. **3** The gases are pushed into a smaller space so they collide more often, which means there are more successful collisions.
for 9.4 **1** It takes place only in the presence of light.
2 carbon dioxide + water $\longrightarrow$ glucose + oxygen **3** Light of different intensities shines on the photographic paper, so the photochemical reaction goes at different rates. The more intense the light is, the darker the shade on the paper.

Questions on section 9
Core curriculum **1 a** Place marble chips and acid in a flask connected to a gas syringe; record the volume of gas produced at regular intervals of time. **b i** increases it **ii** decreases it **iii** increases it **2 a i** a substance that speeds up a chemical reaction, without being used up itself **ii** transition metals
b i Your graph should be a smooth curve, with axes labelled as in the graph on page 66. **ii** Line C will have a steeper curve. (For example, see the two lines on the small graph on page 67.)
iii All the zinc has reacted. **c i** The speed will increase. **ii** The speed will decrease.
Extended curriculum **1 a i** lower concentration of acid: there are fewer acid particles, so there are fewer collisions **ii** greater surface area: more calcium carbonate particles are exposed, so there are more collisions **iii** higher temperature, so particles have more energy and collide more often, with more successful collisions **b ii** From the graph you can find the volume of gas produced in a given period of time. From this you can work out the rate for that period. (See page 66.)
2 a i The total surface area of flour particles is very large, allowing a high number of collisions with oxygen particles.
ii carbohydrate + oxygen $\longrightarrow$ carbon dioxide + water
b The more light that strikes it, the faster the silver bromide decomposes, giving a black deposit of silver. So the uncovered paper is darkest; the paper covered by thick card is white since no light got through. **3 a i** See the graph on page 66. The axes for your graph should be labelled in the same way. **ii** Curve X is steeper since the rate has increased; twice the volume of gas is produced. **iii** The curve is steeper since the rate has increased, but the same volume of gas is produced **b i** more reactant particles present to collide **ii** At higher temperature particles have more energy, so move faster and collide more frequently, and more of the collisions are successful. **c i** glucose, oxygen **ii** chlorophyll
Alternative to practical **1 a** the values for your graph are:

time / min	0	2	4	6	8	10	12
volume / cm³	0	18	30	33	42	45	46

b i at 6 minutes (too low) **ii** 37 or 38 cm³ **2** Using the measuring cylinder, pour some hydrogen peroxide solution into the beaker; test for oxygen at its mouth using glowing splint. Weigh out some manganese(IV) oxide, add to beaker, repeat the test for oxygen. When reaction over, filter the liquid in the beaker; wash the manganese(IV) oxide with distilled water, dry and reweigh. Its mass should be unchanged.

Section 10
Quick checks
for 10.1 **1** It can go both forward and backward. **2** look for the equilibrium sign **3** colour changes from blue to white, and steam comes off **4** add water
5 $CoCl_2.6H_2O$ (s) $\rightleftharpoons$ $CoCl_2$ (s) $+ 6H_2O$ (l)
for 10.2 **1** It is a state where the amounts of reactants and products present do not change, because the forward and backward reactions are going on at exactly the same rate.
2 change temperature; change pressure (for reactions involving gases); change concentration in solutions **3** The forward reaction is exothermic. **4** It speeds up the forward and backward reactions *equally*. **5** lower the temperature, raise the pressure

Questions on section 10
Extended curriculum **1 a** $2NH_3(g) \rightleftharpoons N_2(g) + 3H_2(g)$
b exothermic: negative sign for energy change **c i** Increasing the pressure favours the side of the equation with fewer gas molecules. **ii** The forward reaction is exothermic. **d i** increases it **ii** increases it **e** At lower temperatures the rate is too slow.
2 a i more CO and Cl_2, less $COCl_2$ **ii** Lower pressure favours the side of the equation with more molecules of gas. **b** exothermic: it is favoured by cooling the mixture **3 a** colour change from yellow to orange **b** add alkali **c** OH^- ions remove H^+ ions by neutralisation – so equilibrium shifts to the right to replace them.
4 a i and iii **b** same number of gas molecules on both sides of the equation **c** to increase the rate of reaction
Alternative to practical **1 i** They are the solid in the horizontal tube. **ii** under the solid **b** to condense the water vapour **c** from blue to white **d** Yes, add water to obtain the hydrated crystals.

Section 11
Quick checks
for 11.1 **1** colour change in the compound, from black to red-brown **2** It loses oxygen. **3 a** magnesium **b** carbon dioxide **4** *red*uction and *oxi*dation **5** Yes: carbon dioxide loses oxygen, carbon gains oxygen.
for 11.2 **1 a** $2Na$ (s) $+ Cl_2(g) \rightarrow 2NaCl$ (s) **b** electron transfer from sodium to chlorine **2 a** Sodium is oxidised, chlorine is reduced. **b** $2Na \rightarrow 2Na^+ + 2e^-$; $Cl_2 + 2e^- \rightarrow 2Cl^-$
for 11.3 **1 a** 0 **b** –I **c** –I **2 a** Going along the equation: Mg 0, C +IV, O –II $\rightarrow$ Mg +II, O –II (no change), C 0 **b** There is a change in oxidation states for magnesium and carbon.
c magnesium oxidised, carbon reduced **3 a** +VI **b** +IV **c** +VII
for 11.4 **1 a** magnesium **b** carbon dioxide **2a** Fe 0, Cl_2 0 $\rightarrow$ Fe +III, Cl –I **b i** iron **ii** chlorine **c i** chlorine **ii** iron **3** Both have a strong drive to give up electrons. **4 a** reducing agents **b** oxidising agents **5 a** reduced **b** ii

Questions on Section 11
Core curriculum **1 a** A <u>calcium</u> + oxygen $\rightarrow$ calcium oxide
B <u>carbon monoxide</u> + oxygen $\rightarrow$ carbon dioxide
C iron(III) oxide + <u>carbon monoxide</u> $\rightarrow$ iron + carbon dioxide
b The underlined substances above are being oxidised; the others are reduced.
Extended curriculum **1 a** -I **b** I^- is colourless, I_2 is red-brown **c i** oxidised **ii** $2I^- \rightarrow I_2 + 2e^-$ **d i** –I **ii** change from –I to –II **iii** $2e^-$ **2 a** Acid is present. **b** purple to colourless **c** Fe^{2+}
d when the colour of the solution stays pink **e** $Fe^{2+} \rightarrow Fe^{3+} + e^-$
3 a purple to colourless **b** $2I^- \rightarrow I_2 + 2e^-$

Section 12
Quick checks
for 12.1 **1 a** A acidic, B alkaline, C acidic, D neutral, E alkaline **b** C and E **2** green
for 12.2 **1 a** metal **b** For example:
Mg (s) $+ 2HCl$ (aq) $\rightarrow MgCl_2$ (aq) $+ H_2$ (g)
2 MgO (s) $+ 2HCl$ (aq) $\rightarrow MgCl_2$ (aq) $+ H_2O$ (l) **3 a** carbon dioxide **b** $CaCO_3$ (s) $+ 2HCl$ (aq) $\rightarrow CaCl_2$ (aq) $+ H_2O$ (l) $+ CO_2$ (g)

for 12.3 **1** Heat it with a base such as calcium hydroxide.
2 a The nutrients are not available in a form the plants can use.
b to neutralise the soil's acidity
for 12.4 **1** greater that 7 **2** A strong acid ionises completely, a weak acid ionises only partly. **3** Check table on page 86.
4 a **5** proton **6** neutralisation **7** donors
for 12.5 **1** non-metal **2** any acid **3 a** No, a basic oxide will not react with alkali. **b** Yes, it is an acidic oxide. **4** React it with an acid: ZnO (s) $+ 2HCl$ (aq) $\rightarrow ZnCl_2$ (aq) $+ H_2O$ (l); and also with an alkali: ZnO (s) $+ 2NaOH$ (aq) $\rightarrow Na_2ZnO_2$ (aq) $+ H_2O$ (l). **5** It is neither acidic nor basic. **6** They dissolve in water, giving acids.

Questions on section 12
Core curriculum **1 a** B **b** carbon dioxide **c** A soluble salt forms. **2 a** hydrogen **b i** to make sure all the acid is used up **ii** filtering **c** Evaporate some of the water to give a saturated solution, and allow it to cool. Crystals form and can be filtered off and dried. (See page 9.) **d i** React magnesium oxide, hydroxide or carbonate with sulfuric acid. **ii** The answer depends on your choice in **d i**. **iii** Impurities could harm people.
3 a Acidity increases then decreases. (pH falls then rises.)
b i Saliva is produced gradually, neutralising the sweet's acidity. As more saliva is produced the pH keeps rising. **ii** neutralisation **c i** Carbon dioxide is given off. **ii** Draw a diagram like the first one on page 9, with calcium citrate in the filter paper. **iii** to remove excess lemon juice **iv** See *crystallisation* on page 9.
4 a copper(II) oxide **b** carbon dioxide **c** carbon monoxide **d** water **e** carbon monoxide (or water) **f** calcium oxide, sodium oxide **g** carbon dioxide, phosphorus trioxide
Extended curriculum **1 i** magnesium + sulfuric acid $\rightarrow$ magnesium sulfate + hydrogen **ii** $Li_2O + H_2SO_4 \rightarrow Li_2SO_4 + H_2O$ **iii** H^+ shows an acid is present, so a typical equation is: $CuO + H_2SO_4 \rightarrow CuSO_4 + H_2O$ (or you could use hydrochloric or nitric acid) **iv** sodium carbonate + sulfuric acid $\rightarrow$ sodium sulfate + carbon dioxide + water **b** It accepts a proton, H^+.
c A strong acid is completely ionised in water, a weak one is only partly ionised. **2 a i** The backward reaction is the favoured one, so there is not much product present. **ii** Water donates a proton, methylamine accepts it. **b** less than 12, but over 7; methylamine is only partly ionised so produces fewer hydroxide ions **c i** $CH_3NH_2 + HCl \rightarrow CH_3NH_3Cl$; the salt is methylammonium chloride. **ii** brown/orange precipitate **iii** sodium hydroxide or calcium hydroxide

Section 13
Quick checks
for 13.1 **1 a** magnesium oxide/hydroxide/carbonate and nitric acid **b** zinc/zinc oxide/hydroxide/carbonate and sulfuric acid **c** as for **b** but with hydrochloric acid **2** See the steps on page 93. **3 a** too reactive **b** calcium carbonate/hydroxide/oxide with hydrochloric acid **4** to show when the neutralisation is complete **5** You could choose any sodium, potassium or ammonium salt.
for 13.2 **1** production of an insoluble chemical, during a chemical reaction **2 a** silver nitrate + calcium chloride $\rightarrow$ silver chloride + calcium nitrate
b $2AgNO_3$ (aq) $+ CaCl_2$ (aq) $\rightarrow 2AgCl$ (s) $+ 2NaNO_3$ (aq)
c Ag^+ (aq) $+ Cl^-$ (aq) $\rightarrow AgCl$ (s) **3** calcium and nitrate ions
4 a lead nitrate and sodium or potassium chloride **b** calcium chloride or nitrate and sodium sulfate **c** magnesium chloride, sulfate or nitrate, and sodium carbonate

Questions on section 13
Core curriculum **1 a** These are *insoluble*: calcium and barium carbonates, barium and lead sulfates, silver chloride; the others are soluble. **b i** You could use metal nitrates to provide the metal ions, and sodium salts to provide the negative ions.
ii precipitation **2 a** See page 94. **b** pipette **c** to indicate when neutralisation is complete **d** pink **e** colourless **f** acid: it took only 21 cm³ of acid to neutralise 25 cm³ of alkali **g** potassium chloride

h KOH (aq) + HCl $(aq) \longrightarrow$ KCl (aq) + H_2O (l) **i** evaporate some water to give a saturated solution, cool to obtain crystals, filter
Extended curriculum **1 i** C; sulfuric acid;
ZnO (s) + H_2SO_4 $(aq) \longrightarrow$ $ZnSO_4$ (aq) + H_2O (l)
ii A; hydrochloric acid; KOH (aq) + HCl $(aq) \longrightarrow$ KCl (aq) + H_2O (l)
iii B; potassium or sodium iodide; $Pb(NO_3)_2$ (aq) + 2KCl $(aq) \longrightarrow$ PbI_2 (s) + $2KNO_3$ (aq) **2 a** sodium hydroxide **b** zinc oxide or hydroxide **c** barium chloride or nitrate **d** neutralisation
Alternative to practical **1 a i** spatula **ii** measuring cylinder **iii** tripod **b** more than enough to react with that amount of acid **c** See the first diagram on page 9. **2 a** to make sure all the acid is used up **b** filter **c i** It can dissolve no more solute at that temperature. **ii** Dip a glass rod into it and see if crystals form on the rod. **d** Heating could cause zinc nitrate to decompose (to zinc oxide).

Section 14
Quick checks
for 14.2 **1** A pale blue precipitate forms. If it dissolves in excess ammonia solution giving a deep blue solution, copper(II) ions are present. **2** Ca^{2+}
for 14.3 **1 a** silver nitrate solution **b** a cream precipitate **2** Add hydrochloric acid and barium nitrate. A white precipitate shows sulfate ions are present. **3** Add dilute acid. If a gas is given off test with limewater, to see if it is carbon dioxide. **4 a** aluminium + sodium hydroxide **b** sodium hydroxide
for 14.4 **1** chlorine **2** supports combustion **3** It is poisonous. **4** hydrogen **5** It turns damp red litmus paper blue. **6** A fine white precipitate of calcium carbonate forms.

Questions on section 14
Core curriculum **1 a** See tests on pages 99 and 100. **b** a green precipitate instead of a red-brown one **2 a** Only A will give off a gas (carbon dioxide). **b** See test for nitrate ions on page 100. **c** See tests for Ca^{2+} and Zn^{2+} on page 99. **d** A **3 a** Name any two from chloride, hydrogencarbonate, nitrate, sulfate. **b** calcium **c** 40 mg **d** chloride **e** nitrate **f** slightly alkaline
Alternative to practical **1 a** Q blue /purple, pH 11–14 **b ii** Q no reaction, R fizzes **c** turns limewater milky **e** green precipitate **f** hydrogen **g** carbon dioxide **h** hydrochloric acid **i** weak acid **2 b ii** white precipitate, insoluble in excess **iii** slight precipitate, insoluble in excess **e** weak acids **f** contains Cu^{2+} ions; the smell suggests a compound of ethanoic acid

Section 15
Quick checks
for 15.1 **1** proton number **2** A **3** increases by one each time **4** VII and 0 **5** from non-metal to metal
for 15.2 **1** Group I **2** lithium **3** potassium **4** Reactivity decreases. **5 a** very soft, quite dense, very low melting and boiling points **b** caesium **c** more reactive
for 15.3 **1** Group VII **2** fluorine **3** bromine **4** Reactivity *increases* as you go down Group I but *decreases* as you go down Group VII. **5 a** solid, dark colour **b** iodine **c** less reactive
for 15.4 **1 a** block in the middle **2** high **3** sink **4 a** variable valency (different oxidation states) **b** different colours for different oxidation states **5** nickel
for 15.5 **1** Their atoms have stable outer shells of electrons (2 for He and 8 for others) so do not need to bond to other atoms to gain or lose electrons. **2** Light, unreactive.
for 15.6 **1** from basic to acidic **2** behaves as both an acidic and basic oxide (page 89)

Questions on section 15
Core curriculum **1 a** proton number **b** They have the same relative atomic mass. **c** Group 0 (noble gases) **d** For example Newlands' table has: far fewer elements; no atomic numbers or relative atomic masses; horizontal groups and vertical periods;

no block for transition elements, lanthanides or actinides; Co and Ni in with the halogens; and so on. **2 a** increases **b** around 710 ^{0} – 720 ^{0}C **c** around 0.260 – 0.270 nm **d** quite rapid, but slower than the others **e** All metals conduct electricity and heat, form positive ions, and have metallic bonding.
3 a chlorine + potassium iodide $\longrightarrow$ potassium chloride + iodine **b** Iodine is less reactive than bromine. **4 a** 3 **b** noble gases **c** See page 19. **d** It is inert so will not react with the hot tungsten filament. **5 a** chlorine + potassium bromide $\longrightarrow$ potassium chloride + bromine **b** Iodine is less reactive than chlorine. **c i** gas **ii** black/dark grey **d i** –160 to –190 ^{0}C **ii** 3.0 –3.5 g/dm^3 **e i** 9 **ii** 7
Extended curriculum **1 a** bromine **b** germanium **c** potassium **d** krypton **e** iron **f** bromine **2 a i** BaO **ii** B_2O_3 **b i** S^{2-} **ii** Ga^{3+} **c** Your drawing should show just outer-shell electrons (5 for N, 7 for Cl). N shares a pair of electrons with each Cl. So all four atoms achieve 8 electrons in their outer shells, which is a stable arrangement. **d i** vanadium: higher melting and boiling points, harder, more dense **ii** vanadium: less reactive, coloured compounds, can be used as a catalyst **e i** gas **ii** solid **iii** both poisonous, both react with iron (or other metal)

Section 16
Quick checks
for 16.1 **1 a** can be bent into shape **b** makes a ringing sound when struck **c** can be drawn into wires **2** most react with acids and oxygen **3 a** calcium oxide, Ca^{2+} **b** aluminium chloride, Al^{3+}
for 16.2 **1** rusts easily **2** by turning it into an alloy **3 a** an alloy of copper and zinc **b** is harder than copper, does not corrode **4** See page 113.
for 16.3 **1** hydrogen **2 a** bubbles of gas appear slowly on the magnesium **b** Mg (s) + $2H_2O$ $(g) \longrightarrow$ $Mg(OH)_2$ (aq) + H_2 (g) **c** vigorous reaction, giving magnesium oxide and hydrogen **3** dangerous **4** to show which metal oxides it can reduce **5** iron, copper, silver, gold: all less reactive than Group I metals **6** A protective oxide layer forms on its surface so it appears less reactive than we expect from its position in the reactivity series.
for 16.4 **1 a** Green iron(II) ions replace the blue copper(II) ions. **b** iron + copper(II) sulfate $\longrightarrow$ iron(II) sulfate + copper **2 a** Mg (s) + PbO $(s) \longrightarrow$ MgO (s) + Pb (s) **b** Mg $\longrightarrow$ Mg^{2+} + $2e^-$; Pb^{2+} + $2e^- \longrightarrow$ Pb **c** Mg + $Pb^{2+} \longrightarrow$ Mg^{2+} + Pb
3 a $2KNO_3$ $(s) \xrightarrow{\text{heat}}$ $2KNO_2$ (s) + O_2 (g)
b $2Fe(OH)_3$ $(s) \xrightarrow{\text{heat}}$ Fe_2O_3 (s) + $3H_2O$ (l)
c $CaCO_3$ $(s) \xrightarrow{\text{heat}}$ CaO (s) + CO_2 (g)
for 16.5 **1** They are reactive, and combine with other elements. **2** It is too reactive to be extracted by carbon.
for 16.6 **1** iron(III) oxide + carbon monoxide $\longrightarrow$ iron + carbon dioxide **2** zinc blende, ZnS **3** roast the ore in air to give the oxide **4** Limestone is needed to remove sand (silicon dioxide), which would otherwise be an impurity in the metal.
for 16.7 **1** the iron from the blast furnace; not much demand for it because it cracks easily, due to the high level of impurities in it. **2** by reaction with oxygen, to form carbon monoxide **3** Mild steel is iron with a little carbon, while stainless steel has nickel and chromium added. **4** light but strong, resists corrosion

Questions on section 16
Core curriculum **1 a** aluminium, aircraft bodies; potassium, very soft; platinum, electrodes; iron, extracted from haematite; silver, jewellery **b** fizzing / iron disappears / solution goes green **c i** a metal with other substances mixed with it, to improve on properties **ii** iron **iii** stronger (but too much carbon makes it brittle) **iv** brass, bronze, any other **v** covering with paint, oil, or grease / plating with another metal / sacrificial protection / galvanising **2 a** X, W, Z, Y **b i** W **ii** Y **c** hydrogen **3 a** carbon monoxide **b** Iron(III) oxide loses oxygen. **c** $\underline{3}$C, $\underline{2}$Fe **d i** removes carbon as carbon monoxide, turns phosphorus compounds into

phosphorus oxide **ii** exothermic **iii** It floats on the molten iron. **iv** calcium oxide **v** stronger
Extended curriculum 1 a Oxygen in air would react with metals. **b** sodium, calcium (all more reactive than magnesium) **c** filtration **2 a** eg for roofing, making saucepans, wiring (the moving ions make it malleable and ductile) **b** eg in electrical wiring, saucepan bases (mobile electrons make it a good conductor of heat and electricity) **c** used to make the alloy brass (with copper), and for galvanising iron (see page 119)
3 a $2ZnS (s) + 3O_2 (g) \longrightarrow 2ZnO (s) + 3SO_2 (g)$ **b i** evaporation, condensation **ii** to get as much zinc as possible (equilibrium moves to right) **iii** zinc oxide, carbon dioxide
Alternative to practical 1 a zinc 32°C, iron 16 °C, magnesium 46 °C **b i** magnesium **ii** greatest temperature rise, rapid effervescence **iii** hydrogen

Section 17
Quick checks
for 17.1 1 It dissolves things. **2** to make steam to spin turbines to generate electricity; to cool the steam **3** It turns anhydrous copper(II) sulfate from white to blue, or anhydrous cobalt(II) chloride from blue to pink. **4** no water of crystallisation present **5 a** to remove any remaining solid particles (and filtering with charcoal will remove bad tastes and smells) **b** killed using chlorine
for 17.2 1 mixture, nitrogen, inert, 80%, elements, non-metals **2** four from helium, neon, argon, krypton, xenon, radon **3** They have different boiling points. **4** See page 125.
for 17.3 1 It is poisonous. **2** incomplete combustion of compounds that contain carbon **3** harms plants and fish, corrodes metal and attacks stonework **4 a** converts oxides of nitrogen to nitrogen and oxygen, and carbon monoxide to carbon dioxide **b** It contains a catalyst to speeds up the reactions.
5 $2NO (g) \longrightarrow N_2 (g) + O_2 (g)$ or $2CO (g) + O_2 (g) \longrightarrow 2CO (g)$
for 17.4 1 a two from: burning fuels, respiration, cement production **b** photosynthesis, dissolving in the ocean
2 a Glucose reacts with oxygen in the cells of living things, to give energy. **b** Plants convert carbon dioxide from the air, and water from the soil, to glucose. **3** the burning of fossil fuels in power stations
for 17.5 1 a gas that absorbs heat energy in the atmosphere; carbon dioxide, methane **2** Average air temperatures around the Earth are rising. **3** Burning fossil fuels releases carbon dioxide, a greenhouse gas. **4** The predicted consequences of global warming are serious; all countries will be affected in some way.
for 17.6 1 rusting **2** hydrated iron(III) oxide, $Fe_2O_3.H_2O$
3 keeps air out **4** Any dissolved oxygen has been driven out of the water. **5** $Fe \longrightarrow Fe^{2+} + 2e^-$ or $Fe^{2+} \longrightarrow Fe^{3+} + e^-$ **6** when a more reactive metal is allowed to corrode, to prevent iron from rusting **7** zinc; in two ways: by excluding oxygen and by sacrificial protection

Questions on section 17
Core curriculum 1 a See page 123. **b** It removes any remaining solid particles. **c** to kill bacteria **2 a** nitrogen and oxygen **b i** carbon dioxide **ii** water **iii** $2Cu (s) + O_2 (g) \longrightarrow 2CuO (s)$ **3 a i** Many industries use lead, and in some countries lead compounds are still added to petrol. **ii** See page 126.
b On burning, sulfur forms sulfur dioxide which leads to acid rain; this harms forests and makes lakes acidic, killing fish.
Extended curriculum 1 a loss of electrons **b** hydrogen **c** oxygen, chlorine **d** Oxidation can't take place at a cathode.
e The second metal corrodes, in sacrificial protection; cathodic protection is electrolysis/uses power/produces gases.
Alternative to practical 1 i 4 **ii** 3 **b** There is water and oxygen in both tubes, but less oxygen in tube 3 which contains air. **2 a** flaky, a brown/red colour **b i** because oxygen has been used up **ii** 16.7 % (or 25/150 X 100) **c i** rusting is slower (less iron exposed) **ii** Salt in the water will have no effect (although salt usually speeds up rusting), because in this case the iron is corroded by water vapour, which does not contain salt.

Section 18
Quick checks
for 18.1 1 nitrogen: air; hydrogen: natural gas or cracking hydrocarbons **2** favours the formation of ammonia **3 a** moderate temperature, a catalyst **4** heating an ammonium compound with a strong base **5** provides nitrogen for proteins for growth, and potassium to protect against disease
for 18.2 1 two from: zinc sulfide, natural gas, crude oil, beds of 'native' sulfur **2** normal pressure, moderate temperature (450 °C), catalyst of vanadium(V) oxide **3 a** It acts as a bleach. **b** It kills bacteria.
for 18.3 1 CO_3^{2-} **2** chalk, limestone, marble **3** to remove silicon dioxide (acidic) in the blast furnace; to control acidity in soil; to remove acidic waste gases at power stations
4 Limestone is heated with clay, then gypsum added. **5** It is a hydroxide, quicklime is an oxide. **6 a** removal of sulfur dioxide from flue gases **b** It neutralises sulfur dioxide, which is acidic.

Questions on section 18
Core curriculum 1a give healthier crops, higher yield **b** potassium **c** phosphate **d** neutralisation **e ii** and **iv**
2 a calcium carbonate **b** Acid rain (caused by sulfur dioxide and nitrogen oxides – page 126) attacks limestone, so the statue gets worn away. **c** Damp air causes the iron pin to rust through.
Extended curriculum 1 a natural gas + water **b** Increasing the pressure favours the side of the equation with fewer molecules (ammonia). **c i** iron **ii** gives higher yield of ammonia, since the forward reaction is exothermic (and also reduces fuel costs) **d i** pumped back to the reaction vessel **ii** It has a higher boiling point than nitrogen and hydrogen, so on cooling, it liquifies first. **e i** $2NH_3 (g) + CO_2 (g) \longrightarrow CO(NH_2)_2 (s) + H_2O (l)$
ii It provides only nitrogen. **2 a** colourless gas **b** bleach wood pulp for making paper, food preservative **c i** It is reacted with oxygen or air at 450°C using vanadium(V) oxide as catalyst.
ii $2SO_2 (g) + O_2 (g) \rightleftharpoons 2SO_3 (g)$ **d** ammonium sulfate **e** acid rain **3 i** by burning sulfur in air **ii** It bleaches wood pulp. **iii** It kills bacteria. **b i** decreases **ii** exothermic: as temperature rises yield falls **iii** A lower temperature improves the yield but the rate is too low; a higher temperature increases the rate but gives a poor yield. **iv** It is dissolved in concentrated sulfuric acid to form oleum, which is then carefully mixed with water.

Section 19
Quick checks
for 19.1 1 no – does not contain carbon **2** CH_4, C_4H_{10} **3 a** yes **b** no **c** no
for 19.2 1 separated into groups of compounds with similar boiling points; by fractional distillation **2** the one with 15 **3** bitumen **4** Carbon dioxide is produced when they burn; it is a greenhouse gas and many scientists believe that it is the main cause of global warming. (Not all agree.)

Questions on section 19
Core curriculum 1 a i part of a mixture that has been separated by fractional distillation **ii** fuel gas **iii** On heating, compounds boil off at different temperatures; they are condensed and collected as groups of compounds with a small range of boiling points. **b** See page 140. **2 a i** coal **ii** natural gas **b i** two of: petrol, kerosene, diesel, fuel oil **ii** naphtha, bitumen, lubricating oil **iii** liquid air, and ethanol from the fermentation mixture **3 a** fossil fuels **b** methane **c i** a compound containing only carbon and hydrogen **ii** by fractional distillation **d** naphtha, chemicals; fuel oil, home heating; gasoline, cars; refinery gas, bottled gas; kerosene, aircraft

Section 20
Quick checks
for 20.1 1 An alkene has a double bond. **2** propan-1-ol or propan-2-ol **3 a** alcohol **b** carboxylic acid **4** See propanoic acid,

in upper table on page 143.

for 20.2 1 same functional group **2** by 1 carbon atom at a time **3 a** C_8H_{18} **b** $C_{10}H_{20}$ **c** $C_7H_{15}OH$ **d** $C_5H_{11}COOH$ **4 a** alkene **b** alcohol **c** alkane **5** around 130 –140 °C (actual value is 138 °C)
for 20.3 1 compounds with same formula, but different molecular structure **2** These are the structures:

but-1-ene   but-2-ene   2-methyl propene

3 It has 3 carbons in main chain (like propane), and a branch with 1 carbon (like methane)
for 20.4 1 Each carbon has 4 single bonds **2** unreactive
3 $C_2H_6 (g) + 3\frac{1}{2}O_2 (g) \longrightarrow 2CO_2 (g) + 3H_2O (l)$ (or multiply each number by 2) **4** Hydrogen atoms are replaced by chlorine atoms.
5 $CH_4 (g) + Br_2 (l) \longrightarrow CH_3Br (l) + HBr (g)$ **6** no reaction (no light)
for 20.5 1 C_5H_{10} **2** reactive, due to the double bond
3 bromine water – it goes orange to colourless with alkenes
4 the bromine atoms add on at the double bond
5 poly(propene) **6** only one
for 20.6 1 $C_5H_{11}OH$ **2** OH **3** water molecule adds on, with no other product **4** ethene + water $\longrightarrow$ ethanol
5 $CH_3OH (l) + 1\frac{1}{2}O_2 (g) \longrightarrow CO_2 (g) + 2H_2O (l)$
for 20.7 1 COOH **2** methanoic acid **3** fermentation by bacteria **4** orange $Cr_2O_7{}^{2-}$ ions are reduced to green Cr^{3+} ions
5 The pH of its solution will be under 7, but not very low.
6 by the reaction between a carboxylic acid and an alcohol

Questions on section 20
Core curriculum 1 a Its members have same same functional group so same chemical properties / chain length increases by 1 carbon atom at a time / same general formula / show trend in physical properties **b** ethane, see page 142 **c i** See structure diagram on page 142. **ii** making poly(ethene) and ethanol **iii** See page 143. **iv** $C_2H_4Br_2$ **v** methane **vi** fuel **2 a** carbon dioxide and water **b i** breaking down molecules into smaller ones **ii** heat, catalyst **iii** See structure diagram on page 142. **iv** $C_{12}H_{26}$ **3 a i** C **ii** A **iii** B **iv** D **b i** compound containing only carbon and hydrogen **ii** correct general formula, no double bonds **4 a i** monomers **ii** alkenes **iii** contains C=C double bonds, compound contains only carbon and hydrogen **iv** bromine water changes from orange to colourless when mixed with alkenes, no change with alkanes **b** addition **5 a i** OH **ii** carboxylic acid **iii** CH_3COOH **b** microorganisms
Extended curriculum 1 a methyl ethanoate **b** propanoic acid **c** ethene **2 a** biological catalyst made by living cells **b** reaction of glucose with oxygen in living cells (giving carbon dioxide, water, and energy) **c** Too much heat will kill yeast / denature enzymes. **d** all glucose used up; yeast killed or damaged by ethanol **e** filtration **f** fractional distillation **3 a i** 35 cm³, 40 cm³ **ii** It forms carbon monoxide which can kill (by preventing blood from carrying oxygen). **b i** 1-chlorobutane (or butyl chloride) **ii** UV light **iii** You could show the Cl on a C in the middle of the chain, giving 2-chlorobutane, or show two Cl anywhere on the chain. **c i** See the drawing of poly(propene) on page 155. (Ignore the dotted brackets.) **ii** You could give butan-1-ol or butan-2-ol:

butan-1-ol   butan-2-ol

iii $CH_3-CH_2-CH_2-CH_2-Cl$ or $CH_3-CH_2-CH(Cl)-CH_3$
Alternative to practical 1 a The 'heat' label is in the wrong position, and theres is no water in the trough **b** Bromine water (an aqueous solution of bromine) will go from orange to colourless when added to the tube containing ethene.

Section 21
Quick checks
for 21.2 1 a reaction in which small molecules with a C=C bond add on to each other to form very long-chain molecules, and there is only one product **2** no, no double bond
3 a poly(propenamide);   **b** tetrafluoroethene

for 21.3 1 Two different monomers join; there is always another product – the small molecule eliminated where the monomers join **2 a** and **b** See the circled linkages on the drawings in the second column of the tables on page 156.
3 a a diaminoalkane and a dioylchloride (an alkane with a COCl group at each end of the chain) **b** An N at the end of the diaminoalkane molecule joins to a C at the end of the dioylchloride molecule, eliminating HCl. **4 a** In each, two different monomers join, eliminating a small molecule.
b The types of monomers, the new bonds formed and the molecules eliminated are all different.
for 21.4 1, 2 See page 157. **3** Yes, they would get broken down and disappear.

Questions on section 21
Extended curriculum 1 a and **b** See the second table on page 155. (Ignore the dotted brackets.) **c** polymer **d** monomer – it is unsaturated, with a C=C bond **e** things like plastic bottles, dishes, bowls, water pipes, furniture **f i** cannot be broken down by bacteria **ii** See page 157. **2 b** acrylonitrile **c**

d $nCH_2{=}CHCN \longrightarrow$ ($CH_2{-}CHCN$)$_n$
e Identical monomers add on to each other to form long-chain molecules, and there is only one product.
3 a Two different monomers join to form long-chain molecules; a small molecule is eliminated where they join. **b i** nylon **ii** Terylene **c** See page 156. **d i** a polyester **ii** See the ester linkage on page 156.

Section 22
Quick checks
for 22.1 1 nylon **2** breaks down back to amino acids
for 22.2 1 by a condensation reaction: C on one atom joins to O on another, with the elimination of H_2O **2** See page 160.
for 22.3 1 fats **2** See last row of upper table on page 161.
for 22.4 1 a It breaks down to amino acids through reaction with water. **b** Boil the protein with 6M hydrochloric acid for 24 hours. **2** Heat with dilute hydrochloric acid. **3** Carry out hydrolysis, then use chromatography (with a locating agent).

Questions on section 22
Extended curriculum 1 a ethanoic acid, butan-1-ol
b i See second table on page 156. **ii** fibre to make thread, fabrics for shirts etc **c i** 8 **ii** See page 146. **iii** corn oil
d i 762g **ii** 3 **iii** 3 **2 a** Wood forms carbon dioxide when it burns; soya crop will take in carbon dioxide for photosynthesis, but perhaps less than thick forest. **b i** fats **ii** See page 160.
c i same amide linkage (–CONH–) **ii** *Synthetic polyamide*: just two monomers, one with two –NH_2 groups, the other with two C=O groups. *Proteins*: many different amino acid monomers, each with one –NH_2 and one –C=O group.
3 a i biological catalyst made by living cells **ii** See page 160.
iii chromatography **b i** See page 159. **ii** amino acids
c i propan-1-ol + ethanoic acid $\longrightarrow$ propyl ethanoate + water
ii See the diagram on page 150. **iii** In fat 1 the long chain $C_{17}H_{33}$ contains a double bond; so this fat will decolourise bromine water. But the chain in fat 2 is saturated ($C_{17}H_{35}$).
iv glycerol and sodium salts of fatty acids (or soaps)

Index